# MODELING AND PRICING IN FINANCIAL MARKETS FOR WEATHER DERIVATIVES

# ADVANCED SERIES ON STATISTICAL SCIENCE & APPLIED PROBABILITY

Editor: Ole E. Barndorff-Nielsen

*To view the complete list of the published volumes in the series, please visit:
http://www.worldscientific.com/series/asssap

Advanced Series on

Statistical Science &

Applied Probability

Vol. 17

# MODELING AND PRICING IN FINANCIAL MARKETS FOR WEATHER DERIVATIVES

## Fred Espen Benth

*University of Oslo, Norway*

## Jūratė Šaltytė Benth

*University of Oslo, Norway*

**World Scientific**

NEW JERSEY · LONDON · SINGAPORE · BEIJING · SHANGHAI · HONG KONG · TAIPEI · CHENNAI

*Published by*

World Scientific Publishing Co. Pte. Ltd.

5 Toh Tuck Link, Singapore 596224

*USA office:* 27 Warren Street, Suite 401-402, Hackensack, NJ 07601

*UK office:* 57 Shelton Street, Covent Garden, London WC2H 9HE

**Library of Congress Cataloging-in-Progress Data**
Benth, Fred Espen, 1969–
   Modeling and pricing in financial markets for weather derivatives / by Fred Espen Benth
   (University of Oslo, Norway) & Jūratė Šaltytė Benth (University of Oslo, Norway).
     p. cm. -- (Advanced Series on Statistical Science and Applied Probability ; Vol. 17)
   Includes bibliographical references and index.
   ISBN 978-981-4401-84-5 (hard cover : alk. paper)
   1. Weather derivatives. 2. Stocks--Prices. I. Šaltytė Benth, Jūratė. II. Title.
   HG6052.B46 2012
   332.64'57--dc23

                                          2012026661

**British Library Cataloguing-in-Publication Data**
A catalogue record for this book is available from the British Library.

Printed in Singapore.

Til Julia, smurfen vår

# Preface

There has been a tremendous growth in the weather markets since their beginning around 1997. The organized market for trading weather exists at the Chicago Mercantile Exchange, which launched their first futures and option contracts on temperature in 1999. In the last decade, these markets have also attracted significant attention from academia, with a growing body of literature essentially following three main streams. The first stream focuses on statistical modelling of the time dynamics of weather variables like temperature, rainfall or wind speed. The second stream of research is targeted to pricing and hedging of weather derivatives. Finally, the financial and economical applications of weather derivatives are investigated in the third direction. The current monograph focuses on the first two topics, where the authors have contributed with both new stochastic models for weather variables and developed technology for pricing weather derivatives based on the modern theory of mathematical finance.

The contents of the monograph are to a great extent influenced by our work over the last 10 years in the field of weather derivatives, with particular focus on temperature modelling and futures pricing. The market for temperature derivatives is also currently the most active weather market. Furthermore, we have made contributions to modelling and pricing wind speed derivatives, as well as introducing the concept of hedging spatial risk (or geographical risk, as we call it in this monograph) in weather markets. Lastly, some recent innovations on modelling and pricing precipitation derivatives are presented in the monograph. All this comes in addition to the presentation of relevant work on weather derivatives done by other authors. Most notably, we devote a full chapter on utility-based weather derivatives pricing.

The purpose of this monograph is to present an integrated approach to weather derivatives. We develop stochastic models for the time and space evolution of weather variables on empirical grounds. These are next applied to price weather derivatives using the no-arbitrage theory from mathematical finance. Our methodology provides the fundamental for a correct pricing and risk management in weather markets. Our hope is that the monograph will stimulate further research along these lines in this exciting area of financial engineering where you can buy and sell weather.

In our work on weather derivatives, we are grateful for all the help and fruitful discussions to Andrea Barth, Brenda Lopez Cabrera, Wolfgang Härdle, Paulius Jalinskas, Paul C. Kettler, Jürgen Potthoff and Laura Šaltytė. We especially want to express our gratitude to our friend and collegue Steen Koekebakker, in particular for bringing CARMA to our work on weather. The Lithuanian Meteorological Services are thanked for providing the weather data used in this monograph. Finally, Fred Espen Benth acknowledges the financial support from the project "Managing Weather Risk in Electricity Markets (MAWREM)" funded by the Norwegian Research Council under grant RENERGI 216096.

*Oslo, June 2012*
*Fred Espen Benth and Jūratė Šaltytė Benth*

# Contents

## Weather derivatives      75

# Chapter 1

# Financial markets for weather

By the year 2011, the notional value of the market for weather derivatives was USD11.8 billion, according to the Weather Risk Management Association (WRMA)[1]. The weather markets have grown remarkably in financial strength since the first known weather deals took place in 1996. Nowadays, these markets provide a platform for managing risk exposure in weather variables like temperature, wind and precipitation. The most liquid weather derivatives are based on temperature. In short, these derivatives convert weather into money, where you can profit on bad weather (or good, for that matter).

In this monograph we analyze typical weather derivatives traded in the market, both over-the-counter and as customary assets on exchanges. Our aim is to present a unified approach to the statistical modelling of weather factors like temperature, wind speed and precipitation, and apply these models along with the arbitrage theory of mathematical finance to price weather derivatives. In this first Chapter we present typical weather derivatives and the markets they are traded in, before we move on with the statistical analysis of weather and the pricing of weather derivatives.

## 1.1 The use of weather derivatives

A skiing resort in the Alps is dependent on snow to operate and profit from tourism. A farmer needs fair weather when harvesting the crop in the autumn. A power producer earns money when the weather is hot and air-conditioning is required. A wind mill farm can only operate when there is wind. If weather conditions are unfavourable, on the other hand, the ski

---

[1]see www.wrma.org

1

resort will not attract any tourists, the farmer risks to lose the harvest, the wind mill cannot operate or the demand for power may be low. Too cold or too hot temperatures, no wind or too strong wind, drought or flooding, may seriously harm industry and constitute a major risk to revenues. Weather derivatives are designed to offer a financial tool for hedging this risk.

A farmer, say, can insure the crop against flooding in the harvest season. If there is a flooding, the farmer can claim coverage of the incurred losses based on the insurance contract. However, to make these claims, the farmer must prove that losses are due to the flooding, and this may not be a simple task since also other factors influence the harvest, making it difficult to assess the damage. Weather derivatives, like, for example, derivatives written on rainfall, are an alternative that provides an "objective" way to insure against flooding. Typically, a rainfall derivative is a financial contract that pays the owner money according to an index measuring the amount of rain over a period. For example, the farmer may buy such a contract written on the amount of rain during the harvest season. If there is flooding, the amount of rain will be above a certain threshold, and the contract, properly designed, will pay out money in such a case. The farmer will receive cash, no matter what the actual losses are. The farmer has in effect an insurance against flooding without having to prove any damages. In fact, the farmer will receive money no matter if there is a flooding or not. Of course, the rainfall derivative contract will cost money, which is paid up front at entry.

The profit from a skiing resort is directly linked to the weather during the season. Too warm temperatures destroy snow conditions and harm the profit. To make matters simple, let us suppose that the resort can measure the losses being proportional to the aggregate temperatures $T(t)$, with $t$ being time, above a threshold $c$ in the season, that is,

$$\text{Loss} = a \times \sum_{t \in \text{season}} \max(T(t) - c, 0),$$

for some proportionality constant $a > 0$. It is natural to suppose that the threshold $c$ could be equal to, or be slightly above 0 degrees Celsius (°C). The losses will then be proportional to the number of *cooling-degree days* in the season (see below). Imagine now a temperature derivatives contract which pays the buyer the number of cooling degree days with threshold $c$ over the season measured at the ski resort, in return of a fixed amount $F$. This would in effect be a futures contract on the cooling-degree day index over the season. The ski resort could buy $a$ such contracts, and would in this case receive

$$a \times \sum_{t \in \text{season}} \max(T(t) - c, 0),$$

which covers the temperature-dependent losses. On the other hand, the ski resort must pay the amount $F$. The temperature derivatives contract has in effect *swapped* their stochastically floating loss function with a fixed one.

Consider now the electricity market, for example, in Southern Europe. It is natural to imagine that the profits of a power producer are proportional to temperatures in the summer due to the demand of air-conditioning cooling. The lower temperature, the lower demand and thus less production of power that leads to reduced income. At the same time, prices of power are likely to go down with lower temperatures. The producer can hedge the production by, for example, *selling* futures contracts on the cumulative temperature in the summer period, which would generate a fixed income of money (being the futures price) in return of paying the realized cumulative temperature. If this becomes lower than the futures price, the producer generates an income that could cover the losses. On the other hand, the producer will lose on the futures contract if temperatures become high, but in this case generate income from production. In any case, the producer has locked in the loss generated by temperature to be the fixed futures price, and not the random cumulative temperature.

A retailer in the same market typically have fixed price and volume contracts with the clients, and buys power in the market to honour these. As volume increases with increasing temperatures, the retailer may suffer *losses* incurred by increasing power prices. The retailer may be interested in off-loading the temperature risk by entering long positions in cumulative average temperature futures. A long position in the futures corresponds to buying it, and the retailer will receive the realized cumulative average temperature in return of the fixed futures price. As the retailer on the other hand "pays" the cumulative temperature indirectly in terms of losses, such an investment will make the retailer immune against variations in cumulative temperature. As we see from these two examples, the needs of hedging against temperatures may be opposite for actors in the energy market.

Weather derivatives may prove fruitful for hedging weather risk exposure for outdoor amusement parks, open-air concerts, or soft drink producers, economic activities very different from agriculture or energy. However, nearly all industrial activity is affected by weather in one way or another, so the demand for weather-related hedging tools should be wide. In fact, weather derivatives provide an interesting asset class for speculation and risk diversification for investment funds. As the folklore "London stock markets do not care if it rains in New York" claims, one can apply weather derivatives as an independent asset class in financial investments.

## 1.2     Markets for weather derivatives

The Chicago Mercantile Exchange (CME) organizes a market for weather derivatives. At the CME, futures on temperature, snowfall, hurricanes and rainfall are offered for trade, along with different types of options written on these. Today, CME is the only market for trading in weather derivatives. There have been attempts to set up markets for similar and other weather derivatives. For example, in 2007 the US Futures Exchange (USFE) initiated a market for wind speed derivatives, however the exchange was closed shortly after.

The CME launched its first weather derivatives in 1999, being futures and options on temperature indices measured in several US cities. Nowadays, the weather segment of CME includes futures and options on temperature indices for cities in the US, Canada, Europe, Japan and Australia (we refer to www.cme.com for a complete list). Moreover, the exchange offers derivatives on hurricanes in the Gulf region of the US, snowfall and rainfall derivatives for New York and Boston, and a frost index at Schiphol airport in Amsterdam. There has also been organized trade in weather derivatives at other exhanges for shorter periods, like the Inter Continental Exchange (ICE) and the London International Financial Future and Options Exchange (LIFFE).

### 1.2.1     *Temperature derivatives*

At the CME, temperature futures contracts are settled against three main indices: cooling degree-day (CDD), heating degree-day (HDD) and cumulative average temperature (CAT). The CDD and HDD indices are measured against a benchmark temperature of 65° Fahrenheit (F), or 18°C. Here and in most of this book we shall stick to Celsius as measurement unit for temperature. For a particular day $t$, we define the CDD index as the difference between the average temperature $T(t)$ on that day and the benchmark, whenever this is positive. Otherwise the CDD index is zero.

In mathematical terms, we have

$$\text{CDD}(t) = \max(T(t) - 18, 0). \tag{1.1}$$

The average temperature for a given day $t$ is defined as the mean of the recorded maximum and minimum temperature, that is,

$$T(t) = \frac{T_{\min}(t) + T_{\max}(t)}{2}. \tag{1.2}$$

The CDD index is intended to measure the demand for air-conditioning cooling. The warmer it is, the more cooling is required. For example, for an average temperature of 20°C, the CDD index becomes two, while a recorded temperature of 30°C yields an index value of 12. As we see, the CDD index gives a number which is intimately connected to the volume demand for cooling. Hence, for an energy producer, the higher index value for a given day means more electricity demanded. The index measures this volume, and is the underlying for futures contracts in the summer season.

The HDD index, on the other hand, measures the demand for heating. It is mathematically defined as

$$\text{HDD}(t) = \max(18 - T(t), 0). \tag{1.3}$$

One can imagine that households are putting on heating whenever the temperatures drop below 18°C. For example, a temperature of 0°C will give a HDD index of 18. The CDD and HDD indices are used for US and Australian cities as underlying for the temperature futures (using Fahrenheit as the unit of temperature). The futures are *not* settled on the index value for a particular day, but the *aggregated* index value over an agreed period of time, which we will refer to as the *measurement period*. The measurement periods are typically a week, month, or a longer period like two to seven consecutive months (then referred to as a seasonal strip). A seasonal strip is within the same general season, where *winter* is defined from October through April, and *summer* from April through October. For example, one could imagine a HDD index for New York measured over one week in January. If $\tau_1$ is Monday of that week, and $\tau_2$ is Sunday, then the HDD index for New York will be

$$\text{HDD}(\tau_1, \tau_2) = \sum_{t=\tau_1}^{\tau_2} \max(18 - T(t), 0). \tag{1.4}$$

If the observations for the week in question would be Monday 5°C, Tuesday 6°C, Wednesday 0°C, Thursday 10°C, Friday 12°C, Saturday 19°C and finally Sunday 19°C, the HDD index becomes

$$\text{HDD}(\tau_1, \tau_2) = 39.$$

This would give a measurement of the demand of heating in New York for that given week. The HDD index is the underlying for futures contracts in the winter season.

A futures contract on a CDD or HDD index over a given measurement period is settled against the index value times a cash amount. For the US cities, this cash amount is USD20 per index point. In the above example, the buyer of the HDD futures will receive USD39 × 20, or USD780, in return for the agreed futures price. If the futures contract was entered in December, say, at a futures price of USD750, the buyer would earn a profit of USD30 on this transaction. If the futures price at time of entry was USD800, the buyer would have to pay USD20 effectively.

The CAT index is used as underlying in the summer season for cities in Canada, Europe and Asia. The index is computed over a contracted measurement period $[\tau_1, \tau_2]$ as

$$\text{CAT}(\tau_1, \tau_2) = \sum_{t=\tau_1}^{\tau_2} T(t), \qquad (1.5)$$

that is, the cumulative amount of temperatures from $\tau_1$ to $\tau_2$. The measurement periods are monthly or seasonal strips. The CAT index is substituting the CDD index since in many cities in Europe and Canada the average daily temperature is hardly above 18°C in the summer. The winter index is the HDD as for the US cities. For Canadian cities, the cash amount paid to the buyer of the futures is CAD20, whereas the London futures are settled using the factor GBP20 and the other European locations EUR20. The three cities in Japan covering the Asian area use the average of the CAT index as the underlying, called the Pacific RIM index. It is simply defined as $\text{CAT}(\tau_1, \tau_2)/(\tau_2 - \tau_1)$, and the settlement is based on the Pacific RIM index times YEN2500. All futures are settled immediately after the measurement period has terminated.

On all the temperature futures listed at the CME one can also trade in call and put options. These options are of European-type, meaning that exercise can only take place at a fixed exercise time. A call option on a HDD futures for New York can be used to lock in a fixed futures price at a given time. The option will ensure a maximum futures price given by a strike price $K$. If the futures price is above this strike $K$ at exercise, the option owner will use the contract and enter the futures at the strike price $K$, yielding a profit equal to the difference between the futures and strike prices. The option is abandoned if the futures price is below the strike at exercise, as then it is better to buy the futures in the market instead.

## 1.2.2 Derivatives on wind speed

Generation of wind power has become increasingly important as a renewable energy source in the last decades. Large wind mill farms both on- and off-shore are planned as a substitute for carbon-intensive power production in the EU area. As wind power takes over more and more of the power production, the energy markets will become increasingly more sensitive to the variations in this weather factor.

For example, wind power can only be generated when there is wind, which may create instabilities in the supply of power to the grid. In the German power market EEX, wind generated power has priority in the electricity grid, which creates instances of *negative* power prices as the wind power is fed in at unexpected time periods. Since it is rather costly to ramp down a coal-fired plant for a short time period, it may be better for the operator to simply pay someone to consume the power. On the other hand, the wind power producer faces a large degree of production uncertainty due to the variations in wind speed. The production is dependent on a certain speed of wind at the farm. However, if the wind is too strong, the mills must be closed down in order to avoid damaging the rotors. Clearly, there is a need for risk management tools towards wind uncertainty in the power markets.

As already mentioned, the USFE organized a market for futures settled on wind speed in 2007. The market never came to be, but we discuss its specifications here since this plays the role as a natural set-up of a wind speed derivative market. The USFE applied the so-called Nordix wind speed index as the underlying of their futures contracts. This index was based on the daily average wind speed in two wind farm areas in the US, one in the state of New York, and one in Texas. The two areas were separated into five subareas (two in Texas, three in New York), and futures contracts could be traded on the index in each of these five subareas. In addition, European call and put options were listed on the futures contracts.

The Nordix wind speed index, or the Nordix index from now on, is based on the deviations of the daily wind speed from a 20-year mean. These deviations are then aggregated over a measurement period, being typically a month or a season consisting of consecutive months. Furthermore, it is benchmarked at 100. If we let $W(t)$ be the average daily wind speed measured on day $t$, and $w_{20}(t)$ the mean of the last 20 years' wind speeds for day $t$, the Nordix index measured over the period $[\tau_1, \tau_2]$ is then defined as

$$N(\tau_1, \tau_2) = 100 + \sum_{t=\tau_1}^{\tau_2} (W(t) - w_{20}(t)). \tag{1.6}$$

A futures written on this index pays the buyer the index amount in cash, settled at the end of the measurement period $\tau_2$.

From the definition of the Nordix index, we see that if $N(\tau_1, \tau_2) > 100$, there has been more than average wind in the period, while an index value below 100 means less than average. Hence, if $N(\tau_1, \tau_2) > 100$, a producer of wind power in the farm area has been able to produce *more* than expected, as long as the expectation is based on the mean wind speed over the last 20 years. The producer may face losses if the index falls below 100, implying less than average production. This can be hedged by entering a position in futures on the index, which essentially is equivalent to swapping a floating index with a fixed, or in terms of wind speed, swapping a floating aggregated wind speed against a fixed one.

### 1.2.3    *Precipitation derivatives*

The CME has for some years offered futures on snowfall in US cities. Recently, rainfall futures have been added to the list of precipitation-based weather derivatives traded at this exhchange. The rainfall derivatives are traded for a number of US cities (visit www.cme.com for a complete list). The traded products are futures on a rainfall index measured in each city as well as options on the futures.

The snowfall futures are settled with respect to the CME Snowfall Index, which measures the amount of snow falling in a given measurement period. The period may be a month or a season. At CME one trades in futures for the months November through April, and a season is defined to be two to seven consecutive months (called a seasonal strip, as for temperature derivatives). The buyer of such a futures will receive USD500 times the index value in return for the futures price.

The CME Snowfall Index over a measurement period $[\tau_1, \tau_2]$ is defined as the aggregated daily snowfall in inches over the period. Mathematically, if $S(t)$ is the total amount of snow on day $t$, the Snowfall index will be

$$\mathrm{SF}(\tau_1, \tau_2) = \sum_{t=\tau_1}^{\tau_2} S(t).$$

This index is analogous to the CAT index for temperature.

The settlement of rainfall futures is based on the similar CME Rainfall Index. Instead of the aggregated amount of snow in the measurement period, the amount of rain measured in inches at the location is aggregated instead. The settlement of the futures will be USD500 times the index value. The measurement periods are again months or seasonal strips, in the period March through October.

The CME offers plain vanilla European call and put options written on the snow and rainfall futures. In addition, binary options with an American-style exercise procedure can be traded.

### 1.2.4 *Other weather derivatives*

The CME also offers hurricane and frost index futures, two classes of weather derivatives that we will not analyze in this monograph. However, as they are linked to derivatives on wind and temperature, we briefly mention them here.

Hurricanes can lead to major damages. One may cover the risk for incurred losses from a hurricane by entering hurricane derivatives. These derivatives, being either futures or options, provide an attractive alternative to more traditional damage insurance. When buying an insurance contract, one must prove the claims incurred by the damage from a hurricane, a procedure which may take time and even lead to costly court cases. A hurricane derivative, on the other hand, settles the transfer of money based on an objective index, and as such may lead to payments even if one was lucky and were not hit by the hurricane.

At the CME, hurricane futures are settled against the CME Hurricane Index (CHI). The CHI is a measure for the potential damage incurred by an actual named hurricane making landfall in the US Atlantic Basin. It is dependent on the wind speed and the radius of its force winds. When the hurricane makes a landfall, or dissipitates, the futures is settled by USD1000 per index point. In case the hurricane does not make a landfall, the futures settles at zero.

Call and put options on the CHI are also traded at various strikes. Moreover, the exchange organizes trade in binary options paying a fixed amount if the option is "in" or "at" the money. All the options are cash settled. There exist two more types of hurricane derivatives, one based on the number of hurricanes making a landfall within a season, and another one based on the largest hurricane making a landfall. For both classes, both futures, call and put options and binary options are offered for trade.

The frost index futures is based on frost recorded on weekdays from November through March in Amsterdam. The index is representing the danger of frost on the runway at the Schiphol airport, based on certain levels of recorded temperatures. The futures can be used by airline companies and airport operators to hedge themselves against the financial consequences of a full stop in flying due to slippy runways.

In July 1996, Aquila Energy entered a contract selling electricity to Consolidated Edison Co for the month of August. Part of the deal was a weather clause, that entitled Consolidated Edison Co a rebate on the power if August turned out to be cooler than expected, as measured by the CDD index of August for New York. This transaction is considered to be among the first known weather derivatives deals in the market. Weather quanto options are tightly connected to such a structured deal, a class of derivatives which have gained a lot of attention recently.

An example of a typical weather quanto option can be a call option on some commodity like gas, but where the payment is triggered by certain weather events. In a simple case, one could imagine the option is knocked out by pre-defined levels of HDD or CDD. As gas prices are highly temperature-dependent, such temperature-triggered derivatives are very attractive tools to manage volume (or demand) risk along with price risk. The issuer of such contracts also has both the gas market as well as the weather market to hedge the option. In [Caporin, Preś and Torro (2012)] quanto options on electricity and temperature are priced using Monte Carlo based simulation methods. They consider in particular double call options on degree days and electricity prices. [Benth, Lange and Myklebust (2012)] hedge and price weather quanto options using futures contracts.

In recent years there has been a growing interest in designing so-called weather-indexed insurance contracts, see [Barnett and Mahul (2007)], [Barnett, Barrett and Skees (2008)] and [Skees (2008)]. These contracts are very similar to weather derviatives, as they pay out money to the insured according to an objective index of some weather variable rather than based on claims incurred from actual damages. The main differences from weather derivatives are that the weather-indexed insurance contracts are not traded on an exchange and that they are tailormade to individuals in developing countries.

For example, such a weather-indexed insurance contract may be designed to give a farmer protection against too dry weather that may ruin the crop. Based on measurement of the temperature over a period, say, the farmer will get a payment from the insurance contract. The payment may, for instance, be proportional to a CDD-type index, and in this way give the farmer money if temperatures are too hot. Note that the farmer receives a payment depending on the index, and *not* depending on actual losses. The insurance company, on the other hand, will charge a premium for this kind of protection. The premium will be calculated based on the probabilistic properties of the temperature index, which will be measured at some objective station. In [Taib and Benth (2012)] the price of such temperature-indexed weather insurance contracts is analyzed. Another type of weather-indexed insurance could involve protection against flooding, linking the contract to precipitation at some location.

To make these weather-indexed contracts relevant for farmers in rural areas in Africa, say, the measurement of temperature and precipitation must be done at stations in reasonable distance to the insured. This means that the insurance company must choose measurement stations in areas where there may be significant danger of unreliable data collection. This is a major issue in the design of these contracts. As the stations can rarely be chosen at the location of the insured, there is also a major correlation risk involved. The weather conditions at the farmer's location may only partly be described by the location of the measurement station, and thus the farmer is not covered fully for the weather risk. This is an issue that we will return to in a more general perspective, as this is relevant for all weather derviatives markets. We refer to Sect. 7.3, where *geographical hedging* is analyzed taking spatial risk into account.

## 1.3 A brief outlook of the monograph

This monograph is arranged into two parts: the first part deals with the statistics of weather, where we perform a detailed statistical analysis of the three weather variables, temperature, wind speed and precipitation. Using long time series of observations on temperature and wind speed at different locations in Lithuania, we describe the fundamental properties of the weather dynamics, and propose stochastic models based on autoregressive moving average time series with seasonality. Moreover, we also suggest models taking spatial dependencies into account, leading to spatial-temporal stochastic processes. The stochastic modelling of precipitation is treated separately.

The data set of weather variables from Lithuania is quite extensive. We have long series of daily measurements ranging up to 40 years, which enables us to make a detailed study of the weather dynamics. Lithuania is also a reasonably homogeneous geographical region, and therefore advantageous when studying the spatial dependency structure of weather variables. Situated on the European continent, Lithuania serves as an excellent sample case for the study of weather dynamics with the purpose of pricing financial contracts on temperature, wind speed and precipitation.

In the second part of the monograph, we assess pricing of different weather derivatives using the modern financial mathematics theory of no-arbitrage. Our theoretical approach is based on continuous-time stochastic processes. The time series models from the first part can naturally by formulated as so-called continuous-time autoregressive moving average processes, which we introduce and discuss. Except for precipitation derivatives, we apply the standard stochastic analysis for Brownian motion in analyzing prices and hedges.

The monograph requires a basic knowledge of statistics, probability theory and classical stochastic analysis of Brownian motion. The main core of the monograph is self-contained in the presentation of non-classical theory. There are some notable exceptions:

(1) the use of Lévy processes in modelling stochastic volatility in Sect. 6.1,
(2) the application of Malliavin Calculus in hedging temperature options in Sect. 7.1,
(3) the use of independent increment processes (inhomogeneous Lévy processes) in the modelling of precipitation in Chapter 8,
(4) the application of stochastic control theory in Chapter 9 on the indifference pricing approach to weather derivatives.

The standard reference for Lévy processes with applications in financial markets is [Cont and Tankov (2004)]. The interested reader may consult [Nualart (1995)] for an excellent introduction to the Malliavin Calculus. In Sect. 7.1, we make use of some results from Malliavin Calculus regarding the Malliavin derivative, which will be assumed known. A background in the theory of independent increment processes can be found in [Jacod and Shiryaev (1987)]. A simpler presentation of this class of time-inhomogeneous Lévy processes is found in [Benth, Šaltytė Benth and Koekebakker (2008)], with applications to energy markets. In Chapter 8, we provide the necessary material for following the analysis on precipitation modelling and pricing of precipitation derivatives. [Øksendal (1998)] gives a very motivating background on stochastic control theory relevant to our presentation of utility-based weather derivatives pricing in Chapter 9. In particular, we will use dynamic programming in our analysis, the classical tool for treating such control problems. Dynamic programming leads to a class of non-linear partial differential equations, called Hamilton-Jacobi-Bellman equations.

Throughout the monograph we will use the notion of *forward* and *futures* contracts interchangeably, even though most of the weather derivatives traded are of futures type. We refer to [Hull (2002)] and [Duffie (1992)] for a discussion of the differences and similarities of forward and futures contracts.

We refer to [Jewson and Brix (2005)] for a basic introduction to the various techniques relevant to the analysis of weather derivatives. Together with [Geman (1999)], [Jewson and Brix (2005)] also serves as natural background for understanding the functioning of the weather markets. [Geman and Leonardi (2005)] provide a thorough discussion of the various pricing approaches to weather derivatives, highly relevant for our analysis. Large parts of the material presented in this book is taken from the authors' own work (with collaborators) on weather derivatives. We suggest the interested reader to consult [Benth, Šaltytė Benth and Koekebakker (2008)] for an analysis of modelling and pricing in energy markets, highly relevant for weather. Here a chapter on weather derivatives can be found as well.

For a light approach to the principles of derivatives pricing based on the no-arbitrage theory, the reader is advised to have a look at [Benth (2004)]. A more advanced, but yet reasonably tractable introduction to the modern theory of mathematical finance can be found in [Björk (1998)].

# PART 1

# Statistics of weather

# Chapter 2

# Description of weather data and exploratory analysis

In this Chapter we will describe the weather data used for the empirical analysis throughout the book. The data are typical for what one may obtain from meteorological services worldwide, and our statistical analysis to come will serve as an example on how to deal with these and how to construct and estimate stochastic models for temperature, wind speed and precipitation.

## 2.1 Data

Throughout the book data collected in 20 meteorological stations in Lithuania (Fig. 2.1) will be used. The data base is available through the Lithuanian Hydrometeorological Service in Vilnius, Lithuania (http://www.meteo.lt). We consider three weather parameters, daily average temperature measured in degrees of Celsius (°C), daily average wind speed in meters per second (m/s), and daily precipitation measured in millimeters (mm). In most of the stations, observations on these three meteorological variables are available in electronic form since the beginning of 1961. However, in order to have time series of equal length for a specific variable in all stations of interest, we choose a later starting point in time for our analysis.

Four measurement stations (Dūkštas, Lazdijai, Palanga and Trakų Vokė) will be excluded from the analysis because of too many missing observations or because the station in question is a close neighbour of another station (Trakų Vokė, for instance, is nearby Vilnius station). Missing observations in other stations were substituted for each station separately by the average of observations made at the same day in the last five years. The percentage of missing values is low (0.2% in two stations (Laukuva

17

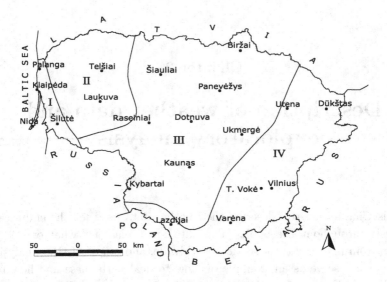

Fig. 2.1    Map of Lithuania with measurement stations. T.Vokė is abbreviation for Trakų Vokė.

and Utena) for daily average temperature, 0.0009% in Kybartai and 0.5% in Laukuva and Utena for wind speed, and 0.009% in Vilnius and 0.3% in Laukuva and Utena for precipitation), and the comparison of imputed and original data sets shows that the imputation method has a negligible influence on data sets.

We proceed now with an exploratory analysis of the three weather variables at each spatial location separately. When illustrating our results graphically, we mainly use only Vilnius station. Due to the simple surface topology of Lithuania the pattern in other stations is similar to that in Vilnius.

## 2.2   Temperature

As a starting point for daily average temperature records we choose 1 June 1964. Daily average temperatures are ranging until 31 August 2004 resulting in 14,692 observations. In Fig. 2.2, we plot a snapshot of temperature data for the last five years in Vilnius station, where we notably observe a clear seasonal pattern in the data.

Some simple descriptive statistics for all 16 stations are given in Ta-

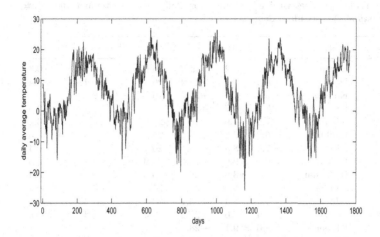

Fig. 2.2   Time series of daily average temperatures for Vilnius. A snapshot of last five years of observations.

ble 2.1. The variation in different characteristics of temperatures is not big among the stations. The values of skewness and kurtosis, small, but significantly different from zero, indicate that temperature might be non-normally distributed. Note that all values of skewness and kurtosis are negative, suggesting left skewness and a less peaky distribution than the normal.

Analysis of histograms in all stations confirms the conclusions reached above. In Fig. 2.3, we present the histogram for Vilnius. Clearly, it has a left skewness and no clear peak which is contradictory to normality. In fact, the distribution seems to be bimodal, an implication of winter and summer seasons with cold and warm weather, respectively. We also check for the presence of autocorrelation in the temperatures. In Fig. 2.4, the autocorrelation function (ACF) for temperature is plotted. The strong seasonal variation in the values of the autocorrelations is a clear sign of seasonality in data.

## 2.3   Wind

We choose 1 January 1977 as a starting date for the analysis of wind speed time series. Data series continue until 31 December 2007, resulting in 11,315 observations. A snapshot of last five years of wind speed in Vilnius is plotted in Fig. 2.5. Wind speed exhibits an annual seasonality with stronger wind

Table 2.1 Descriptive statistics for daily average temperature ('Std' stands for standard deviation).

| Station | Mean | Std | Min | Max | Skewness | Kurtosis |
|---|---|---|---|---|---|---|
| Biržai | 6.3 | 9.2 | −29.9 | 26.9 | −0.4 | −0.3 |
| Dotnuva | 6.6 | 9.0 | −28.9 | 26.5 | −0.4 | −0.4 |
| Kaunas | 6.8 | 8.9 | −26.7 | 26.7 | −0.3 | −0.4 |
| Kybartai | 7.2 | 8.7 | −26.7 | 29.1 | −0.4 | −0.3 |
| Klaipėda | 7.5 | 7.9 | −26.4 | 27.3 | −0.3 | −0.4 |
| Laukuva | 6.0 | 8.6 | −28.8 | 25.8 | −0.3 | −0.5 |
| Nida | 7.6 | 8.3 | −25.1 | 26.7 | −0.3 | −0.4 |
| Panevėžys | 6.6 | 9.0 | −29.2 | 26.9 | −0.4 | −0.3 |
| Raseiniai | 6.3 | 8.8 | −28.2 | 26.5 | −0.3 | −0.4 |
| Šiauliai | 6.5 | 8.9 | −29.7 | 27.3 | −0.3 | −0.4 |
| Šilutė | 7.2 | 8.4 | −26.1 | 26.8 | −0.3 | −0.3 |
| Telšiai | 6.4 | 8.6 | −29.6 | 27.4 | −0.3 | −0.5 |
| Ukmergė | 6.6 | 9.1 | −28.4 | 28.2 | −0.4 | −0.3 |
| Utena | 6.2 | 9.2 | −29.2 | 27.7 | −0.4 | −0.3 |
| Varėna | 6.5 | 9.1 | −27.7 | 28.0 | −0.4 | −0.2 |
| Vilnius | 6.2 | 9.3 | −28.3 | 27.4 | −0.3 | −0.5 |

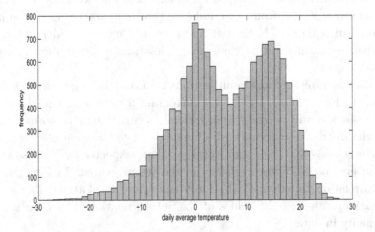

Fig. 2.3 Histogram of daily average temperatures for Vilnius.

speed and larger variations experienced in cold seasons than in warm ones. By analyzing time series in other locations, we observed that the seasonal pattern is more apparent and less noisy for inland stations compared to stations in coastal areas.

The descriptive statistics of the wind speed are presented in Table 2.2.

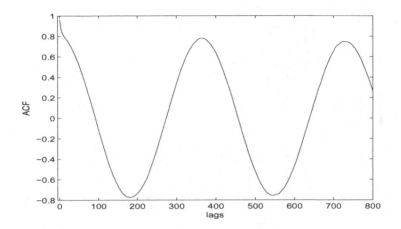

Fig. 2.4 Autocorrelation function of temperatures for Vilnius.

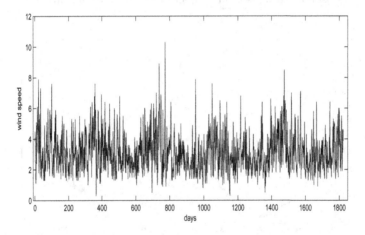

Fig. 2.5 Time series of wind speed for Vilnius. A snapshot of last five years of observations.

The average wind speed varies from 2.4 to 4.7 m/s from station to station with standard deviations being relatively stable over the region. Kurtosis and skewness are all positive and significantly different from zero. The histograms for wind speed in all stations (see Fig. 2.6 for illustration) are strongly right-skewed, suggesting that a proper transformation of data might symmetrize them well. In Fig. 2.7, the ACF for wind speed is plotted,

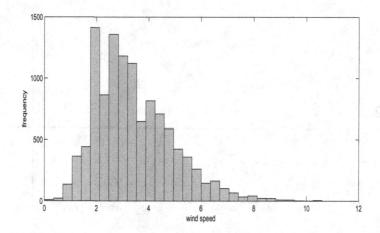

Fig. 2.6    Histogram of wind speed for Vilnius.

where we observe strong seasonal effects and autocorrelations.

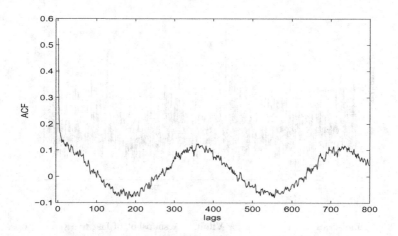

Fig. 2.7    Autocorrelation function of wind speed for Vilnius.

## 2.4    Precipitation

Also for precipitation records January 1, 1977 is chosen as the starting date in this study. Data series continue until 31 August 2004, resulting in

Table 2.2   Descriptive statistics for wind speed ('Std' stands for standard deviation).

| Station | Mean | Std | Min | Max | Skewness | Kurtosis |
|---------|------|-----|-----|-----|----------|----------|
| Biržai | 3.4 | 1.5 | 0.1 | 12.0 | 0.8 | 0.9 |
| Dotnuva | 2.7 | 1.5 | 0.1 | 14.4 | 1.5 | 3.5 |
| Kaunas | 3.8 | 1.7 | 0.3 | 13.1 | 0.8 | 0.8 |
| Kybartai | 3.2 | 1.7 | 0.0 | 19.0 | 1.3 | 2.6 |
| Klaipėda | 4.7 | 2.5 | 0.5 | 22.8 | 1.2 | 2.4 |
| Laukuva | 3.5 | 2.0 | 0.1 | 18.8 | 1.6 | 3.7 |
| Nida | 4.5 | 2.3 | 0.1 | 18.5 | 1.0 | 1.2 |
| Panevėžys | 3.3 | 1.4 | 0.0 | 11.0 | 0.7 | 0.7 |
| Raseiniai | 3.8 | 1.5 | 0.1 | 13.3 | 0.8 | 1.1 |
| Šiauliai | 2.8 | 1.2 | 0.0 | 9.3 | 0.7 | 0.8 |
| Šilutė | 3.8 | 1.9 | 0.0 | 16.3 | 1.2 | 2.2 |
| Telšiai | 3.1 | 1.3 | 0.0 | 12.6 | 1.0 | 2.0 |
| Ukmergė | 3.6 | 1.9 | 0.0 | 16.0 | 1.2 | 2.5 |
| Utena | 2.6 | 1.3 | 0.0 | 9.0 | 0.9 | 1.0 |
| Varėna | 2.4 | 1.2 | 0.0 | 8.5 | 0.8 | 0.7 |
| Vilnius | 3.4 | 1.5 | 0.0 | 10.6 | 0.9 | 0.8 |

10,098 observations. As we can see from a snapshot of last five years of precipitation in Vilnius in Fig. 2.8, precipitation is a highly varying phe-

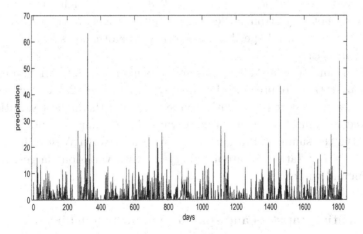

Fig. 2.8   Time series of precipitation for Vilnius. A snapshot of last five years of observations.

nomena. No clear pattern is seen from the time series plot. The histogram in Fig. 2.9, however, demonstrates an extreme right-skewness in data (skewness coefficient varies from 4.1 to 5.4 from station to station), caused by many days without precipitation (corresponding to a measurement of 0 mm). Because of the same reason the kurtosis is also extraordinary high,

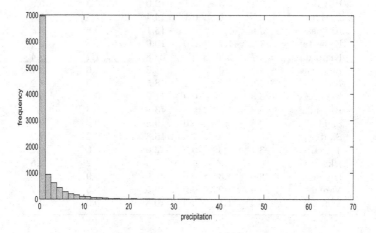

Fig. 2.9    Histogram of precipitation for Vilnius.

and varies from 27.3 to 53.4 in different stations. Some seasonal pattern is observed through the ACF (see Fig. 2.10), even though the autocorrelations are not very strong. Clearly, precipitation behaves differently than temperature and wind speed do, and other type of models will be needed in modelling it.

The means, standard deviations and maximum values (minimum is zero mm in all stations) of precipitation are presented in Table 2.3 for each season separately. The most precipitation on average and the highest variation is observed in summer and/or autumn, with the maximum of precipitation reached in the summer. Stations in the West and North-West of Lithuania get most precipitation all around the year, with variations largest in the autumn and winter.

## 2.5    Spatial statistics and spatial-temporal modelling

To explore spatial properties of data, we calculated the correlations for each pair of time series. For temperatures, all correlations were very close to +1, indication that the temperature is reasonably stable over Lithuania. The correlations for wind speed are somewhat weaker, but all well

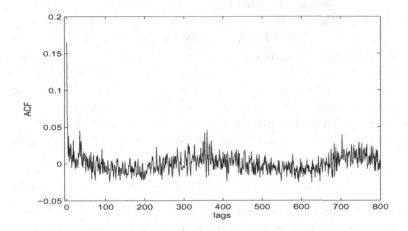

Fig. 2.10　Autocorrelation function of precipitation for Vilnius.

Table 2.3　Descriptive statistics for precipitation (W=winter, Sp=spring, Su=summer, A=autumn, 'Std' stands for standard deviation).

| Station | Mean | | | | Std | | | | Max | | | |
|---|---|---|---|---|---|---|---|---|---|---|---|---|
| | W | Sp | Su | A | W | Sp | Su | A | W | Sp | Su | A |
| Biržai | 1.4 | 1.4 | 2.4 | 1.9 | 2.4 | 3.1 | 5.7 | 3.6 | 18.8 | 35.0 | 84.7 | 37.9 |
| Dotnuva | 1.2 | 1.2 | 2.1 | 1.5 | 2.2 | 3.0 | 4.9 | 3.2 | 29.9 | 58.4 | 64.4 | 36.6 |
| Kaunas | 1.4 | 1.4 | 2.4 | 1.7 | 2.4 | 2.9 | 5.3 | 3.7 | 22.5 | 24.3 | 73.4 | 45.3 |
| Kybartai | 1.3 | 1.3 | 2.4 | 1.6 | 2.4 | 2.9 | 5.4 | 3.5 | 19.9 | 40.2 | 67.6 | 43.6 |
| Klaipėda | 1.9 | 1.3 | 2.3 | 2.9 | 3.1 | 3.0 | 5.5 | 4.9 | 26.9 | 28.4 | 73.9 | 32.8 |
| Laukuva | 2.1 | 1.6 | 2.7 | 2.7 | 3.5 | 3.4 | 6.2 | 5.0 | 31.6 | 37.0 | 84.7 | 48.2 |
| Nida | 1.7 | 1.3 | 2.3 | 2.7 | 3.0 | 2.9 | 5.3 | 4.9 | 39.2 | 40.3 | 72.6 | 45.1 |
| Panevėžys | 1.3 | 1.4 | 2.3 | 1.7 | 2.1 | 3.3 | 5.3 | 3.4 | 18.6 | 61.9 | 86.2 | 34.5 |
| Raseiniai | 1.6 | 1.5 | 2.6 | 2.1 | 2.5 | 3.1 | 5.3 | 3.8 | 24.4 | 37.4 | 72.3 | 36.5 |
| Šiauliai | 1.3 | 1.3 | 2.4 | 1.9 | 2.1 | 3.3 | 5.4 | 3.7 | 19.7 | 48.4 | 63.2 | 43.3 |
| Šilutė | 1.8 | 1.4 | 2.8 | 2.9 | 2.9 | 3.1 | 6.4 | 5.0 | 19.9 | 37.8 | 77.3 | 48.1 |
| Telšiai | 2.1 | 1.5 | 2.7 | 2.7 | 3.6 | 3.4 | 6.6 | 5.0 | 34.6 | 36.1 | 103.8 | 42.8 |
| Ukmergė | 1.4 | 1.4 | 2.4 | 1.9 | 2.4 | 3.2 | 5.2 | 3.7 | 26.2 | 28.5 | 57.2 | 33.7 |
| Utena | 1.5 | 1.5 | 2.6 | 1.9 | 2.4 | 3.4 | 5.7 | 3.7 | 16.3 | 45.0 | 78.4 | 39.8 |
| Varėna | 1.6 | 1.6 | 2.5 | 1.9 | 2.6 | 3.4 | 5.4 | 4.1 | 24.3 | 64.7 | 60.7 | 50.5 |
| Vilnius | 1.6 | 1.5 | 2.5 | 1.9 | 2.5 | 3.4 | 5.5 | 4.1 | 31.0 | 62.5 | 64.3 | 55.8 |

above 0.5. Weakest correlations between stations were observed for precipitation data, with the smallest value down to 0.24 and the largest one equal to 0.78. High correlations for temperature and wind speed are likely due to relatively short distances in Lithuania, where the longest distance

between two stations is approximately 292km. Moreover, temperature and wind speed in Lithuania vary in a quite similar manner on long time scale. Precipitation, on the other hand, is highly variable and a more local meteorological phenomenon, which results in much lower spatial correlations than those for temperature or wind speed. As expected, all correlations were clearly dependent on the distance between stations. For all three meteorological variables, observations in stations within shorter distances were more correlated than observations in stations geographically farer apart.

Spatial-temporal data can be considered as a realization of a random field

$$\left\{ Z(\mathbf{s}; t) : \mathbf{s} \in D \subset \mathbb{R}^d, t \in T \subset [0, \infty) \right\}, \qquad (2.1)$$

where $\mathbf{s}$ and $t$ define spatial and temporal coordinates, respectively. Usually, $d = 2$ or 3. Hence, the observations are placed into a combined space-time framework. Data can be regarded as either a multiple time series or as several realizations of a spatial random field with time as replication index. From a statistical point of view, it is often assumed that a random field $\{Z(\mathbf{s}; t)\}$ (or its increments of certain order) are second-order stationary in space and time.

It is important to distinguish between the coordinates in time and space. Data in time has a clear ordering, which is not common for observations in space. In addition, distance units are different in space and time and cannot be directly compared in a physical sense. Moreover, the phenomena of interest are often periodic in time, whereas they are not in space. This makes modelling in the space-time domain to a non-trivial task.

[Kyriakidis and Journel (1999)] present a thorough review of spatial-temporal models and point out two approaches for modelling of spatial-temporal distributions. One approach decomposes a spatial-temporal random field $Z(\mathbf{s}; t)$ into a mean component $\mu(\mathbf{s}; t)$ modelling the trend (smooth variability), and residual component $\varepsilon(\mathbf{s}; t)$ modelling the higher frequency fluctuations around the trend in both space and time. Such a decomposition of $Z(\mathbf{s}; t)$ can be represented mathematically as

$$Z(\mathbf{s}; t) = \mu(\mathbf{s}; t) + \varepsilon(\mathbf{s}; t),$$

where $\mu(\mathbf{s}; t)$ is a deterministic or stochastic function of the space and time coordinates. The residual field $\left\{ \epsilon(\mathbf{s}; t) : \mathbf{s} \in D \subset \mathbb{R}^d, t \in [0, \infty) \right\}$ is a zero-mean (and usually stationary) spatial-temporal random field. Sometimes the spatial-temporal mean component and/or residuals field can be decomposed into a sum or product of the single functions of spatial and temporal components. The assumptions of separability between and/or stationarity in space and time might greatly simplify the modelling and estimation of the spatial-temporal random field.

The covariance structure is the main object in modelling both the spatial and temporal components. A separability assumption may be reasonable for small scale processes or made out of convenience, especially when dealing with large number of observations. Separable covariance models, a sum or product of purely spatial and purely temporal covariances, are commonly used, with one of the first applications in [Rodriguez-Iturbe and Mejia (1974)] (see also [Høst, Omre and Switzer (1995)], [Solna and Switzer (1996)]). In [Mardia and Goodall (1993)] a survey on spatial-temporal random fields with separable covariance functions can be found. [Gneiting, Genton and Guttorp (2007)] review advances in the literature on spatial-temporal covariance functions. A shortcoming, however, is that separable models arise from two separate processes (one temporal and one spatial) that act independently from each other, implying that the spatial behaviour of the field $\{Z(\mathbf{s}; t)\}$ is considered to be the same at all time instances. In a similar way, no change of the temporal pattern of $\{Z(\mathbf{s}; t)\}$ from one spatial location to another can be assumed. In other words, separable models do not take into account the interaction between space and time. Moreover, the multiplicative covariance models are based on the assumption that for any two spatial locations (any pair of time points), the cross-covariance function has the same shape regardless of the displacement of the locations (time points). For the additive model, covariance matrices of certain configurations of spatial-temporal data are singular, being an undesirable property in kriging, for example. For more details on the shortcomings of separable models see [Rouhani and Myers (1990)], [Myers and Journel (1990)] and [Kyriakidis and Journel (1999)]. In [Fuentes (2006)], a method to test for the separability of spatial-temporal process is suggested.

Development of non-separable covariance functions started with metric models, where spatial and temporal units were converted into a common metric (see [Dimitrakopoulos (1994)]). [Deutsch and Journel (1998)] suggested the use of three-dimensional zonal anisotropy techniques. Classes of non-separable spatial-temporal stationary covariance functions have been proposed by [Cressie and Huang (1999)], [De Iaco, Myers and Posa (2002)], [Gneiting (2002)], [Ma (2003a)], [Fernandez-Aviles, Montero and Mateu (2011)] and others. However, as [Snepvangers, Heuvelink and Huisman (2003)] point out, even though such functions are mathematically correct, they often lack physical support and are somewhat artificial.

Often there is little reason to assume spatial and/or temporal stationarity of the covariance function. Non-stationary spatial-temporal covariance functions rely generally on treatment of semivariogram parameters being themselves Gaussian random fields and are discussed in, for example, [Kyriakidis and Journel (1999)]. [Fuentes, Chen, and Davis (2007)] introduce a general and flexible parametric class of spatial-temporal covariance models, that allows for lack of stationarity and separability by using a spectral representation of the process. [Ma (2003b)] demonstrates how to derive non-stationary spatial-temporal covariance functions via spatial-temporal covariances and intrinsically stationary variograms. [Bruno et al. (2009)] propose an approach for modelling the case where non-separability arises from temporal non-stationarity. Estimation of non-stationary spatial covariance structure from spatial-temporal data is presented in [Nott and Dunsmuir (2002)].

In another approach for modelling the spatial-temporal field, as suggested in [Kyriakidis and Journel (1999)], two model subclasses can be defined. Models in the first subclass treat the spatial-temporal random function $\{Z(\mathbf{s}; t)\}$ as a collection of a finite number of temporally correlated spatial functions $\{Z(\mathbf{s})\}$. The models in the second subclass view the random function $\{Z(\mathbf{s}; t)\}$ as a collection of a finite number of spatially correlated time series $\{Z(t)\}$. The choice of the subclass depends on which domain (spatial or temporal) contains more data points.

One of the first attempts to approach spatial-temporal modelling is done by [Cliff and Ord (1975)], by incorporating spatial autoregressive (AR), moving average (MA), and regressive components into a so-called SARMAR model. [Pfeifer and Deutsch (1980)] propose a spatial-temporal autoregressive moving average (STARMA) models. STARMA models may effectively and simultaneously capture the continuity in space and time. However, the identification and estimation procedures of STARMA models are quite complex, as noticed by [Niu and Tiao (1995)], who use a simple spatial-temporal AR (STAR) process to model satellite ozone data for trend assessment. Moreover, STARMA models fail to take into account nonlinear behaviour like, for example, unusual jumps in rainfall data (see [Park and Heo (2009)]).

[Huang and Cressie (1996)] apply a spatial-temporal model to snow water equivalent data. They decompose the observed process into a sum of unobserved state process and a white-noise process. The temporal dependence in the unobserved state process is modelled as an AR process, and predictions and prediction errors for the state process are obtained through a Kalman filter algorithm. Kalman filtering is an attractive approach in their application since it enables the state variable to be updated once new observations are available. As pointed out by [Cressie and Wikle (2002)], the most typical spatial-temporal models include deterministic spatial-temporal trend plus various spatial, temporal and spatial-temporal random variations. Such an approach requires an explicit specification of joint spatial-temporal covariance structure of the process of interest.

[Handcock and Wallis (1994)] use a traditional kriging approach within a Bayesian framework in order to develop a random field model for the mean temperature over the region in the northern United States. The authors use the static areal quantities as a basis for estimating temperature shifts by the historical record. A Bayesian approach has the advantage that the extra variability due to unknown parameters is automatically taken into account through the prior distribution on them. [Cressie (1994)] points out in a discussion of the paper by [Handcock and Wallis (1994)] that a Bayesian approach to prediction in a spatial-temporal setting can be presented in the context of state-space modelling, and that a Kalman filter incorporating space and time could be useful.

Sometimes, the trend estimation is of interest as well. In [Kyriakidis and Journel (1999)], different trend models are separated into deterministic and stochastic models. As possible deterministic spatial-trend models, [Dimitrakopoulos and Luo (1997)] suggest three general forms, traditional polynomial functions, Fourier expressions, or the mixture of the two. If there are covariates available, a trend component can be well represented by a certain regression model like, for example, in [Carroll et al. (1997)]. If the regressors in the regression model can be considered as random, one can model the trend by a stochastic function (stochastic trend model).

## 2.6 Stochastic weather modelling – literature overview

We review some of the relevant literature on stochastic modelling of weather variables. Particular attention is paid to time series processes for marginal modelling of weather in a specific location. Although we give a rather extensive overview of the existing literature, the intention is not to be exhaustive. We have a focus on those models which are most relevant for the financial applications in this book.

## 2.6.1    *Temperature*

[Tol (1996)] proposes to use a generalized AR conditional heteroskedastic (GARCH) model for daily temperature measurements. In such a model the conditional variance of the model depends linearly on the conditional variances of previous observations and on the previous prediction errors. Tol's motivation for choosing the GARCH model is based on the observation that the predictability of meteorological variables is not constant but shows regular variations. He estimates the proposed model on 30 years of daily temperature records in De Bilti, The Netherlands. [Franses, Neele and van Dijk (2001)] propose to use a so-called QGARCH (quadratic GARCH) model allowing for asymmetry in the impact of innovations on the conditional variance. They estimate the model on the same data set as in [Tol (1996)].

[Caballero, Jewson and Brix (2002)] note that the classical Box-Jenkins approach to time series modelling may result in the underestimation of the variance of seasonal means. They demonstrate that the problem may be overcome by applying a fractional AR integrated MA (ARFIMA) model and estimate it on daily temperatures observed in Central England for 222 years (sic!).

[Dornier and Querel (2000)] suggest a general Ornstein-Uhlenbeck dynamics with time-dependent variance for time series of temperature in Chicago. However, they only consider empirically a constant specification of variance. [Alaton, Djehiche and Stillberger (2002)] use a similar model for data collected from Bromma, Sweden, with a time-dependent variance. They smooth it out on monthly data to obtain a constant variance over each month. [Brody, Syroka and Zervos (2002)] also consider an Ornstein-Uhlenbeck model, however, with a fractional Brownian motion driving the stochastics. [Cao and Wei (2004)] propose the time series approach for modelling the temperature, which results in an ARMA (autoregressive moving average) model with periodic variance. [Campbell and Diebold (2005)] model the daily average temperature in a number of US cities by an AR time series with a seasonal AR conditional heteroskedastic (ARCH) type dynamics for the residuals. [Oetoma and Stevenson (2005)] and [Svec and Stevenson (2007)] compare several models (some of them mentioned above) in terms of their performance in forecasting weather indices, without being able to show that one of them outperforms the others. [Papazian and Skiadopoulos (2010)] compare the model of [Campbell and Diebold (2005)] and a modified model of [Benth and Šaltytė-Benth (2007)] with three alternative models by testing the out-of-sample forecasting performance. Their comparison also ended inconclusive regarding the identification of the best performing model. Yet another comparison of several models for temperature dynamics can be found in [Schiller, Seidler and Wimmer (2010)].

[Solna and Switzer (1996)] build up a model for estimating the temporal trend for an evolving spatial field and estimate it on monthly temperature averages in the stepp region in eastern Europe at 24 monitoring sites. In [Šaltytė Benth, Benth and Jalinskas (2007)], the model in [Benth and Šaltytė-Benth (2005, 2007)] for time series of temperatures estimated on data from, respectively, Norway and Sweden, is extended to a spatial-temporal case and estimated on data collected in 16 meteorological stations in Lithuania. Here, the times series are first modelled at each spatial location separately. Later, the spatial correlation is estimated on the residuals. [Im, Rathouz and Frederick (2009)] analyze 20 years of daily temperature data at 11 stations in Chicago metropolitan region by a spatial-temporal model in a way similar to that in [Šaltytė Benth, Benth and Jalinskas (2007)]. They first build a time series model, containing a long-term trend, annual and semiannual harmonics, different physical covariates and an ARMA process at each spatial location. Next they model the residuals as independent and identically distributed replicates of a spatial process with a covariance structure from the Matérn class. A similar statistical approach was also used by [Li et al. (1999)] to model particulate matter in Vancouver.

## 2.6.2 *Wind*

There is a large body of literature available on the analysis of wind speed time series. The classical statistical models for wind speed are based on Weibull, Rayleigh or lognormal distributions (see, for example, [Corotis, Sigl and Klein (1978)], [Garcia et al. (1998)], [Lun and Lam (2000)], [Celik (2003)]). An implicit assumption in this type of models is the independence of wind speed records, which is not true in practice. Another, more natural way to describe the statistical features of wind speed time dynamics is using ARMA time series models (see, for instance, [Brown, Katz and Murphy (1984)], [Rehman and Halawani (1994)], [Castino, Festa and Ratto (1998)], [Torres et al. (2005)], [Brett and Tuller (1991)] and [Martin, Cremades and Santabarbara (1999)]). A linear time-varying AR process is used by [Huang and Chalabi (1995)] to model and forecast hourly wind speed at Herstmonceux (England). [Bouette et al. (2006)] model univariate time series of wind speed by generalized ARMA process. [Tol (1997)] uses GARCH process to model the daily mean wind speed at Shearwater (East Coast of Canada). [Šaltytė Benth and Benth (2010)] use time series decomposition approach in modelling the wind speed time series in New York.

[Şahin and Şen (2004)] propose a trigonometric point cumulative semivariogram concept for deciding on a spatial dependence function and its use for regional prediction. [Haslett and Raftery (1989)] describe the spatial-temporal structure of wind speed in Ireland at 12 meteorological stations. The authors base their inference on deseasonalization, kriging, and fractional ARMA modelling. The main purpose of their modelling is to evaluate the average power output to be expected in the long term from a wind turbine at a given spatial location with few observations on wind speed available. [Haslett and Raftery (1989)] first estimate and subtract seasonality from the data. They next model the residuals by a mean function and an ARFIMA process at each spatial location separately and assume that the resulting residuals are spatially correlated. Spatial dependencies are then modelled by an exponential correlation function.

This way of modelling spatial-temporal fields was also explored by [Loader and Switzer (1989)] when modelling monthly rainfall acidity data from a monitoring network in the eastern part of the United States. The same approach was used by [Šaltytė Benth and Šaltytė (2011)], where the time series model in [Šaltytė Benth and Benth (2010)] is generalized to a spatial-temporal case. [Denison, Dellaportas and Mallick (2001)] propose to use a completely nonlinear Bayesian multivariate regression spline model for prediction of wind speed at a new spatial location with only short run of data available. They estimate and compare their model with the model of [Haslett and Raftery (1989)] for data from the island of Crete in Greece.

[Yan et al. (2002)] apply a generalized linear model for modelling daily wind speed time series in Northwestern Europe. [Cripps et al. (2005)] use a Bayesian hierarchical model to predict the surface wind field one day ahead within Sydney Harbour. [Ailliot, Monbet and Prevosto (2006)] propose a Markov switching AR model for modelling the spatial-temporal evolution of wind speed. [Chandler and Bate (2007)] use the properties of independence estimating equations to adjust the "independence" log-likelihood function in the presence of clustering and apply such adjustments to the modelling of wind speed in Europe and temperatures in the UK.

### 2.6.3 Precipitation

[Roldan and Woolhiser (1982)] compare two models, a first-order Markov chain and an alternating renewal process, describing the occurrence of sequences of wet and dry days using daily data from five National Weather Service stations in the US. In [Woolhiser and Roldan (1982)], different models for the distribution of the amount of daily precipitation are compared for the same data set. [Mimkou (1983)] studies the seasonality of the Markov chain order for modelling daily precipitation occurrences for seven stations in Greece. [Rodriguez-Iturbe, Cox and Isham (1987)] introduce a stochastic model for rainfall intensity at a fixed point in space, and assume that each storm event consists of a cluster of a random number of rain cells and arrives according to a Poisson process. The suggested model is extended in [Rodriguez-Iturbe, Cox and Isham (1988)].

[Onof, Faulkner and Wheater (1996)] examine methodologies for the simulation of storms and propose two new approaches, one combining point profiles of storms extracted from long rainfall records and another using a stochastic rainfall model based upon the Poisson process to generate data. The authors use the data from a small catchment in Surrey, England. A rainfall model is divided into two sub-models in [Todini, and di Bacco (1997)], where the first one models the total number of rainfall spells within a window of time by a Pólya process, while the second one, conditional on the first, describes the total quantity of rainfall in the time window, given the number of rainfall spells.

One of the first models for rainfall combining the information from time and space is proposed by [Rodriguez-Iturbe and Mejia (1974)]. The authors consider the rainfall process as multidimensional random field with a separable correlation function. They estimate long-term mean areal rainfall in a homogeneous area of 30,000 km$^2$ with 26 rain gauges in central Venezuela. [Cox, and Isham (1988)] generalize the results in the paper by [Rodriguez-Iturbe, Cox and Isham (1987)] to a spatial-temporal model of rainfall. [Dalezios and Adamowski (1995)] develop a spatial-temporal precipitation model from the general class of STARMA processes and apply it to the precipitation time series from 11 rain gauge stations spatially located within the Grand River watershed in Ontario, Canada. A stochastic spatial-temporal model of rainfall in which the arrival times of rain cells occur in a clustered point process is presented by [Cowpertwait (1995)] and fitted to the rainfall data from six spatial locations in the Thames basin, UK. [Guenni and Hutchinson (1998)] fit a simplified model of [Rodriguez-

Iturbe, Cox and Isham (1987)] to daily rainfall recorded at each of 102 locations from South-East Queensland, Australia, and carry out the spatial estimation of the parameters in locations without rainfall records by using thin plane smoothing splines. [Northorp (1998)] develops a model which is a spatial analogue of the point process-based models used to represent the temporal rainfall process at a single spatial location by [Rodriguez-Iturbe, Cox and Isham (1987, 1988)] and which is a generalization of the spatial-temporal model suggested by [Cox, and Isham (1988)]. [Wheater et al. (2000)] develop a model of spatial rainfall, with respect to both full spatial-temporal modelling and the multi-site problem of modelling correlated rain gauge data. The authors estimate the discussed models on data collected through the HYREX project in UK. [Brown et al. (2001)] build a dynamic regression model at each time series separately, and then estimate the spatial-temporal covariance structure of the minimum mean-square error predictors of the dynamic regression coefficients in order to guide the formulation of an integrated spatial-temporal model for the region of interest. The final spatial-temporal model is fitted by likelihood-based methods to the data from the weather radar station located in Lancashire, England. In [Velarde, Migon and Pereira (2004)], a spatial conditional AR model is considered where precipitation is modelled by a two-stage process. In particular, the model is dealing with the problem of non-negative variables which have a significant point mass at zero level. Here the spatial interpolation and prediction of weekly rainfall levels in the central region in Brazil is considered in a Bayesian context. [Zhang and Switzer (2007)] propose an event-based model, which is continuous in two-dimensional space and time, describing regional scale, ground-observed storms by a Boolean random field of rain patches. The authors estimate their model using hourly data at eight rain gauges in Alabama, US. [Park and Heo (2009)] model the monthly and regional variation of rainfall fields in South Korea using a seasonal spatial-temporal bilinear model.

# Chapter 3

# Spatial-temporal stochastic modelling of weather

Our main goal in this Chapter is to build stochastic models for weather variables that are simple statistically but yet sufficiently sophisticated to explain the basic stylized facts of the weather dynamics. We aim at building a model for each of the three variables; temperature, wind speed and precipitation. As it turns out, the wind speed and temperature dynamics share some statistical properties that makes it possible to apply the same kind of stochastic processes to model their dynamics, namely ARMA time series. Precipitation has different properties, and the model class for this weather variable will be chosen separately. Our aim is to construct models which allow for an explicit derivation of prices of various financial weather derivatives contracts. To reach this goal, time series models that can be formulated in continuous time are particularly attractive. The models suggested in this Chapter satisfy this property. In Chapter 4 we discuss the continuous-time analogues of the stochastic models introduced here.

We applied Matlab v7.0, SPSS v16 and Excel in all the statistical analyses of this Chapter.

## 3.1 The modelling approach

We have at hand daily data on temperature, wind speed and precipitation from 16 meteorological stations in Lithuania over 30 years or longer, which will be used to understand the spatial-temporal structure of weather parameters over the region. With this extent of temporal data compared to the relatively few spatial locations, it is natural to start with modelling the time series behaviour marginally. We therefore first analyze time series at each spatial location separately. The approach where spatial and temporal components are modelled separately has earlier been used by [Haslett

35

and Raftery (1989)] for modelling the wind speed structure in Ireland at 12 meteorological stations. [Loader and Switzer (1989)] analyze monthly rainfall acidity data in the United States by also first modelling the temporal dynamics at each spatial location separately and next fitting the spatial correlation function on the residuals.

As seen from the exploratory analysis in Chapter 2, the temperature and wind speed have similar marginal behaviour with a distinct seasonal pattern, a possible time trend, and a relatively symmetric distribution. However, the level of variation around the mean at each spatial location is of noticeable difference for these two weather parameters. Wind speed is a rather noisy parameter and therefore more complicated to model compared to temperature. In addition, wind speed values cannot be negative, whereas temperature obviously can. Precipitation can also take only nonnegative values, but has a mass point at zero (from periods of no rain) and a highly skewed distribution. Therefore, precipitation requires a different type of stochastic model.

The time series analysis of temperature and wind speed presented here follows closely the analysis in a number of papers published by the authors with collaborators. A time series model for temperature was first published in [Benth and Šaltytė-Benth (2005)] and [Benth and Šaltytė-Benth (2007)] for, respectively, Norwegian and Swedish data. It was then generalized to the spatial-temporal case in [Šaltytė Benth, Benth and Jalinskas (2007)] for Lithuanian temperature. Wind speed time series were first analyzed in [Šaltytė Benth and Benth (2010)] and estimated for New York observations. Later, the marginal time series model for wind speed was generalized to the spatial-temporal model in [Šaltytė Benth and Šaltytė (2011)], and applied to Lithuanian data.

Briefly, our modelling approach is as follows. We first analyze each time series separately independently of spatial information. In view of the many discussions on climate changes, a first step in our data analysis is to check for presence of a (linear) trend. This might be particularly relevant for temperature records. Naturally, temperature and wind speed vary with the season, as seen from the exploratory data analysis. Therefore, a component modelling seasonal variations is a natural part of the model. The remaining cyclic variations left in the data are modelled with an ARMA process. Different model components are eliminated from the data in a stepwise fashion, and autocorrelations and histograms are examined at each step to assure that the residuals produced by the final model follow the properties of a white noise process. After eliminating trend, seasonal component and

ARMA effects, we experience that the residuals often still do not exhibit properties characteristic of a white noise process, calling for more sophisticated modelling. The residuals show clear signs of heteroskedasticity, which we propose to model by a truncated Fourier series.

The parameter estimates in the time series models turn out to be reasonably stable over space, supporting the idea of separating time and space components. Our time series model at each spatial location yields residuals where most of the time dependency has been removed. By assuming that the residuals come from a Gaussian random field, we model the spatial correlation structure based on empirical correlations between stations. We observe that the spatial correlations have a very clear structure leading to simple and precise modelling. Moreover, a relatively simple trend-surface model describes well the spatial behaviour of the time series parameters over the region.

As we already mentioned, the time series of precipitation at each spatial location carries a different character than those of temperature and wind speed. Clear clustering of values around zero (days without rain) suggests to separate the dynamics into 'rainy' and 'not rainy' days. The intensity and amount of precipitation on rainy days can be modelled as two separate processes, blending deterministic seasonality effects with random variations. A multiplicative model is desirable in such a setup to ensure positivity.

The detailed modelling of the three weather variables are presented in the next sections, which are organized as follows. In the coming section we state a spatial-temporal model for temperature and wind speed. Next, the time series modelling at a single spatial location independently of the spatial information is presented in detail. Finally, we model spatial correlation function on time-independent residuals and combine all parameters from time series analysis into a spatial model. A model for precipitation time series is presented in the last section.

## 3.2 Spatial-temporal model for temperature and wind speed

We consider spatial-temporal data to be a realization of a random field as defined in (2.1). As discussed earlier, it is typical to decompose a spatial-temporal field $Z(\mathbf{s};t)$ into a mean component $\mu(\mathbf{s};t)$ modelling the trend, and a residual component $\varepsilon(\mathbf{s};t)$ modelling the fluctuations around the trend in both space and time. Hence, the decomposition of the spatial-temporal random field $Z(\mathbf{s};t)$ can be written as

$$Z(\mathbf{s};t) = \mu(\mathbf{s};t) + \varepsilon(\mathbf{s};t), \qquad (3.1)$$

where $\mu(\mathbf{s};t)$ is a deterministic function of space and time coordinates. We suppose it to be given by

$$\mu(\mathbf{s};t) = \Lambda(\mathbf{s};t) + \sum_{i=1}^{p_\mathbf{s}} \alpha_i(\mathbf{s})(Z(\mathbf{s};t-i) - \Lambda(\mathbf{s};t-i)) - \sum_{j=1}^{q_\mathbf{s}} \beta_j(\mathbf{s})\varepsilon(\mathbf{s};t-j), \quad (3.2)$$

where $\Lambda(\mathbf{s};t)$ describes the (linear) trend and seasonality in terms of sine and/or cosine functions in $Z(\mathbf{s};t)$, and $\alpha_i(\mathbf{s})$ and $\beta_j(\mathbf{s})$ are parameters of an ARMA$(p_\mathbf{s}, q_\mathbf{s})$ process at the spatial location $\mathbf{s} \in D$. The parameters $\alpha_i(\mathbf{s})$ and $\beta_j(\mathbf{s})$ are assumed to be non-random, even though they can in general be time-dependent. As we do not have any strong evidence empirically for such a time-dependency, we suppose them to be only spatially varying. Furthermore, we factorize the residual field into

$$\varepsilon(\mathbf{s};t) = \sigma(\mathbf{s};t)\epsilon(\mathbf{s};t), \qquad (3.3)$$

where $\sigma(\mathbf{s};t)$ is a non-random seasonal function, satisfying the annual periodicity condition

$$\sigma(\mathbf{s};t) = \sigma(\mathbf{s};t+365) \qquad (3.4)$$

for any time $t$. Assume that

$$\{\epsilon(\mathbf{s};t) : \mathbf{s} \in D \subset R^2, t \in [0,\infty)\} \qquad (3.5)$$

is a zero-mean stationary Gaussian spatial-temporal random field which is independent in time and has a spatial correlation function defined by the parametric model

$$\mathrm{corr}\,(\epsilon(\mathbf{s};t), \epsilon(\mathbf{s}+\mathbf{h_s};t)) = \rho(\mathbf{h_s};\theta_\mathbf{s}) \qquad (3.6)$$

for all $\mathbf{s}, \mathbf{s}+\mathbf{h_s} \in D$, $\mathbf{h}$ being a distance between two spatial locations in $D$.

It is reasonable to expect that for measurements in close neighbouring spatial locations there is some cross-correlation in time and space in the short term. However, the data are aggregated to daily averages and it is very likely that on such a time scale these correlations blur out. We therefore assume that the cross-correlation in time and space is equal to zero. We also note that the residuals $\varepsilon(\mathbf{s};t)$ are uncorrelated but dependent in time and correlated in space. Therefore, the model of the spatial-temporal covariance function for $\varepsilon(\mathbf{s};t)$ defined in (3.3) is

$$C(\mathbf{s}, \mathbf{s}+\mathbf{h_s};\theta) = \rho(\mathbf{h_s};\theta_\mathbf{s})\Sigma_\mathbf{s}(\theta_t)\Sigma_{\mathbf{s}+\mathbf{h_s}}(\theta_t), \qquad (3.7)$$

where $\Sigma_\mathbf{s}(\theta_t)$ is a diagonal variance matrix at the location $\mathbf{s} \in D$. We denote by $\theta = (\theta_\mathbf{s};\theta_t)' \in \Theta$ the $l \times 1$ parameter vector, where $\Theta$ is an open subset of $R^l$. Here the notation $\mathbf{x}'$ means transposing of the vector $\mathbf{x}$ (or matrix, if $\mathbf{x}$ is such).

### 3.2.1 *Marginal modelling of temperature and wind speed*

Denote by $Z_k(t)$ a time series at the spatial location $s_k \in D$, $k = 1, ..., n$, $t = 1, ..., T$. Then

$$Z_k(t) = \mu_k(t) + \varepsilon_k(t), \tag{3.8}$$

where $\mu_k(t)$ and $\varepsilon_k(t)$ denote the mean and residual process at the time moment $t = 1, ..., T$ at the spatial location $s_k \in D$, respectively. Here

$$\mu_k(t) = \Lambda_k(t) + \sum_{i=1}^{p_k} \alpha_i^k (Z_k(t-i) - \Lambda_k(t-i)) - \sum_{j=1}^{q_k} \beta_j^k \varepsilon_k(t-j), \tag{3.9}$$

with $\alpha_i^k$ and $\beta_j^k$ being parameters of an ARMA$(p_k, q_k)$ process and

$$\Lambda_k(t) = a_0^k + a_1^k t + \sum_{l=1}^{L_\Lambda} \left( a_{2l}^k \cos(2l\pi t/365) + a_{2l+1}^k \sin(2l\pi t/365) \right). \tag{3.10}$$

We assume that the residual process $\varepsilon(t)$ is of the following form

$$\varepsilon_k(t) = \sigma_k(t)\epsilon_k(t),$$

where $\sigma_k(t)$ is a (possibly) time-dependent volatility function, and $\epsilon_k(t)$ is a zero-mean temporally independent Gaussian random process with standard deviation equal to one. Following the analysis of temperature and wind speed in [Benth and Šaltytė-Benth (2005, 2007)], [Šaltytė Benth, Benth and Jalinskas (2007)], [Šaltytė Benth and Benth (2010)], and [Šaltytė Benth and Šaltytė (2011)], one way to model the variance $\sigma_k^2(t)$ is to fit a truncated Fourier series

$$\sigma_k^2(t) = b_1 + \sum_{l=1}^{L_\sigma} \left( b_{2l}^k \cos(2l\pi t/365) + b_{2l+1}^k \sin(2l\pi t/365) \right) \tag{3.11}$$

to the empirical daily variance of the residuals. The alternative is to use the daily empirical variance of residuals to approach the properties of a white noise process. It was also observed in [Šaltytė Benth and Benth (2012)] that it might be necessary with an even more complex model, including a GARCH-type stochastic volatility process.

Remark that the model parameters can in general depend on the location $s_k \in D$, $k = 1, ..., n$. The shape of $S_k(t)$ and/or $\sigma_k^2(t)$, and the order of an ARMA$(p_k, q_k)$ process depend on the phenomenon to be modelled. Since the variations in wind speed, for example, are much more complex than in temperature, the order of the ARMA$(p_k, q_k)$ process is naturally expected to be higher for wind speed. The seasonal function can be affected in a similar manner as well. We will therefore discuss the model estimation process for each weather parameter separately.

### 3.2.2    *Spatial modelling of temperature and wind speed*

In order to specify the spatial-temporal random field $Z(\mathbf{s};t)$, we define a spatial model for each of the parameters of the time series models. For this purpose, we fit a trend-surface model (see [Ripley (1981)])

$$P^{\eta}(x,y) = \lambda_{00}^{\eta} + \lambda_{10}^{\eta}x + \lambda_{01}^{\eta}y + \lambda_{20}^{\eta}x^2 + \lambda_{11}^{\eta}xy + \lambda_{02}^{\eta}y^2 + \ldots + \lambda_{rw}^{\eta}x^r y^w \quad (3.12)$$

to each set of parameters. Here, $x$ and $y$ are $n \times 1$ vectors of metric coordinates of stations, and $\eta$ indicates the set of parameters and takes values $a_l$, $l = 0, ..., L_{\Lambda}$, $\alpha_i$, $i = 1, ..., p_k$, $\beta_j$, $j = 1, ..., q_k$, and $b_l = 1, ..., L_{\sigma}$. The sum $r + w$ represents the order of the trend-surface, first order or linear for $r + w = 1$, second order or quadratic for $r + w = 2$, third order or cubic for $r + w = 3$, and so on.

The residuals produced by the time series models exhibit white noise properties. We therefore calculate the empirical correlations between the residuals for all pairs of stations, giving us $n(n-1)/2$ values. The spatial correlation function is then fitted to the empirical correlations.

### 3.2.3    *Estimation of the marginal temperature model*

For modelling the temperature dynamics at each spatial location, we use a time series decomposition approach. Here, the time series is decomposed into different components, like trend, seasonality, ARMA process and residual term. All these components appear in the data simultaneously. By estimating and eliminating different components of the time series step-by-step and examining the residuals at each step, we get a good insight into the structure of the data, and are in this way likely to reach a rather precise model.

#### 3.2.3.1    *Trend*

A number of empirical studies on temperature data shows that there is a clear increase in temperature (see, for example, [Alaton, Djehiche and Stillberger (2002)] for Swedish temperatures and [Campbell and Diebold (2005)] for various US locations), which is in line with the discussions on global warming (see [Rassmusson et al. (1993)] and [Handcock and Wallis (1994)]). We choose to model the trend as a linear function of time (see equation (3.10)), with a long-term average temperature $a_0$, and the trend $a_1 t$ ensuring stationarity in temperature time series. Even though the linear trend might be a simplification when a longer time series of temperature is considered, a constant trend from year to year seems to be validated in our case.

We run a simple linear regression model at each station. The obtained slopes are all significantly different from zero and approximately equal to 0.0001, meaning that the temperature has increased in Lithuania by about 1.5°C in the period 1964-2004. The positive trend corresponds to the increase in the global mean temperature, which has risen over the past 100 years by about 0.6°C and is now increasing more rapidly (see the report [IPCC AR4 WG1 (2007)] by the Intergovernmental Panel on Climate Change). The intercepts vary in the interval [5.46, 6.91] and are all significantly different from zero at the level of 1%.

After performing this trend analysis, we proceed to the estimation of the function (3.10), where all components are estimated simultaneously.

### 3.2.3.2 *Seasonal component*

The deterministic function $S(t)$ in (3.10) is modelling the trend and seasonality of temperature. Analysis of temperature suggests that a choice of a truncated Fourier series gives a sufficiently flexible class of functions to describe the temperature. Such a sum of trigonometric functions explains the seasonal variations in the temperature, appearing as low temperatures in the winter and high in the summer. According to the standard statistical significance tests, we conclude that a rather low order truncated Fourier series explains the seasonal variations in the temperature well. The estimates of the fitted function (3.10) are reported in Table 3.1 for each station. We notice, that a different approach to modelling the seasonal function for temperature using a local smoothing technique was proposed and applied in [Härdle et al. (2010)]. Yet another approach using wavelet analysis for seasonality modelling was suggested by [Zapranis and Alexandridis (2008)].

### 3.2.3.3 *ARMA process*

We next eliminate the estimated trend and seasonal effects from the temperature and examine the ACF, plotted in Fig. 3.1. We clearly managed to remove most of the seasonality from the temperature by eliminating the seasonal component. However, there is still a strong memory effect present in the deseasonalized data. The partial ACF (PACF) shown in Fig. 3.2 suggests that an AR(3) process should be sufficient for explaining the remaining autocorrelations. The estimated parameters of the AR(3) process are reported in Table 3.1; all estimates are significant at the 1% level. For

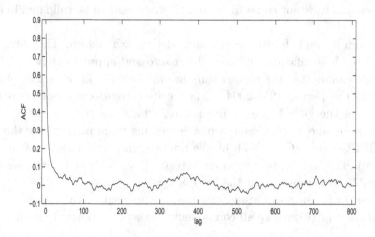

Fig. 3.1    ACF of deseasonalized temperatures for Vilnius.

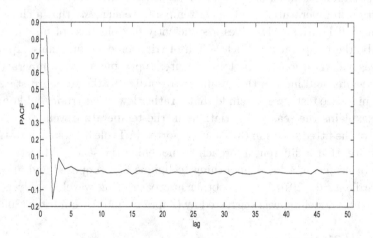

Fig. 3.2    PACF of deseasonalized temperature for Vilnius.

comparison, we also estimated an AR(4) process on deseasonalized data, and found that this model fitted the data slightly better. However, the differences in parameter values were minor. Moreover, the difference in the values for the Akaikes' Information Criteria (AIC) for the two models were negligible. Hence, we prefer to use the simpler of the two to model the residuals.

The ACF of the residuals obtained after eliminating the seasonal component and the AR(3) process (see Fig. 3.3) shows no autocorrelations present in the data. The histogram of these residuals is depicted in Fig. 3.4

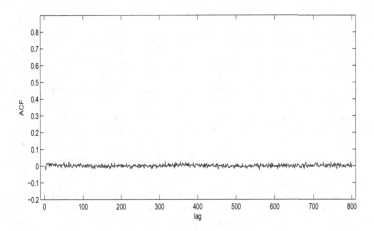

Fig. 3.3   ACF of residuals obtained after eliminating seasonal component and AR(3) process for Vilnius.

and seems to be symmetric. However, the Kolmogorov-Smirnov (KS) test

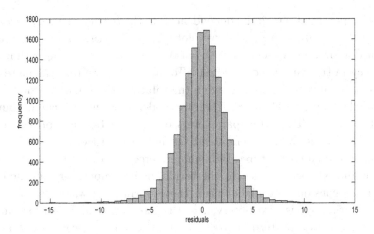

Fig. 3.4   Histogram of the residuals obtained after eliminating seasonal component and AR(3) process for Vilnius.

does not support the hypothesis about normality. In fact, the KS statistics vary from 4.04 to 6.30 and $p < 0.001$ for all stations, with Vilnius being no exception.

Table 3.1    Fitted parameters of the linear trend, seasonal component and AR(3) process.

| Stations | $a_0^k$ | $a_1^k (\times 10^{-5})$ | $a_2^k$ | $a_3^k$ | $\alpha_1^k$ | $\alpha_2^k$ | $\alpha_3^k$ |
|---|---|---|---|---|---|---|---|
| Biržai | 5.45 | 10.52 | 8.20 | 7.89 | 0.947 | −0.226 | 0.093 |
| Dotnuva | 5.72 | 11.10 | 8.07 | 7.82 | 0.950 | −0.223 | 0.091 |
| Kaunas | 6.17 | 7.86 | 8.04 | 7.75 | 0.954 | −0.229 | 0.092 |
| Kybartai | 6.38 | 9.66 | 7.67 | 7.60 | 0.951 | −0.233 | 0.096 |
| Klaipėda | 6.69 | 9.58 | 6.11 | 7.75 | 0.922 | −0.213 | 0.105 |
| Laukuva | 5.39 | 8.11 | 7.57 | 7.66 | 0.959 | −0.228 | 0.089 |
| Nida | 6.83 | 10.44 | 6.46 | 8.14 | 0.934 | −0.209 | 0.109 |
| Panevėžys | 5.73 | 10.42 | 8.10 | 7.78 | 0.945 | −0.222 | 0.090 |
| Raseiniai | 5.63 | 8.29 | 7.83 | 7.70 | 0.954 | −0.227 | 0.092 |
| Šiauliai | 5.71 | 9.45 | 7.81 | 7.85 | 0.961 | −0.243 | 0.102 |
| Šilutė | 6.42 | 9.18 | 7.17 | 7.52 | 0.929 | −0.204 | 0.090 |
| Telšiai | 5.50 | 10.70 | 7.44 | 7.66 | 0.976 | −0.253 | 0.100 |
| Ukmergė | 5.84 | 9.58 | 8.15 | 7.82 | 0.938 | −0.222 | 0.092 |
| Utena | 5.46 | 9.58 | 8.24 | 7.85 | 0.935 | −0.220 | 0.091 |
| Varėna | 5.84 | 8.08 | 8.16 | 7.75 | 0.903 | −0.185 | 0.078 |
| Vilnius | 5.44 | 8.84 | 8.48 | 7.93 | 0.966 | −0.243 | 0.090 |

### 3.2.3.4    *Residuals*

To detect potential remaining dependencies in the residuals, we analyze the ACF of the *squares*. In Fig. 3.5 we plot the ACF of the squared residuals obtained above, and observe a clear seasonal pattern, indicating a time dependency in the variance of residuals. To model this, we first calculate the daily empirical variance by averaging the values of the squared residuals of a particular day over all years. Then we model it by the truncated Fourier function (3.11). The fitted parameters of this function are presented in Table 3.2. In Fig. 3.6, the empirical variance $\sigma_k^2(t)$ together with the fitted one is presented for three Lithuanian meteorological stations (Vilnius, Šiauliai and Klaipėda). Clearly, the variations in temperatures in the cold season are considerably higher than those during the warm season. Increased variability is also observed in the late spring and early summer. This characteristic pattern is observed in all stations, however, with some differences related to the geographical location of the stations. The function is flattest in inland stations (for example, Vilnius (Fig. 3.6, left)), gets

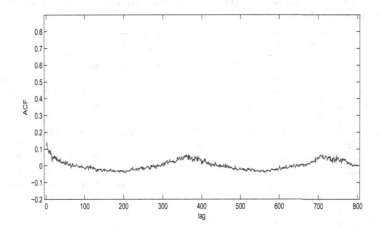

Fig. 3.5   The ACF of squared residuals obtained after eliminating seasonal component and AR(3) process for Vilnius.

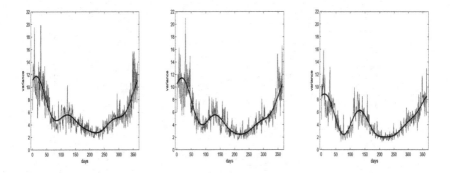

Fig. 3.6   The empirical and fitted variance for temperature residuals for Vilnius (left), Šiauliai (middle), and Klaipėda (right).

more pronounced when moving towards coast (Šiauliai (Fig. 3.6, middle)), and becomes quite wavy on the coast (Klaipėda (Fig. 3.6, right)).

The seasonal dependency in the variance is eliminated by dividing the residuals by the square root of the fitted variance. The ACF and histogram for final residuals are presented in, respectively, Fig. 3.7 and Fig. 3.8. The plots basically show that we are left with zero-mean uncorrelated in time noise which is close to normally distributed. The KS test statistics vary from 2.75 to 4.51 over the different stations, and are highly significant

Table 3.2  Fitted parameters of $\sigma_k^2(t)$. Asterisk marks parameters not significantly different from zero at the level of 5%.

| Stations | $b_1^k$ | $b_2^k$ | $b_3^k$ | $b_4^k$ | $b_5^k$ | $b_6^k$ | $b_7^k$ | $b_8^k$ | $b_9^k$ |
|---|---|---|---|---|---|---|---|---|---|
| Biržai | 5.83 | 1.07 | 3.75 | 0.36 | 1.59 | 1.28 | 0.72 | 0.46 | −0.11* |
| Dotnuva | 5.54 | 0.94 | 3.31 | 0.18* | 1.49 | 1.12 | 0.70 | 0.53 | 0.04* |
| Kaunas | 5.40 | 0.88 | 2.84 | 0.07* | 1.29 | 0.92 | 0.76 | 0.52 | 0.11* |
| Kybartai | 5.63 | 0.96 | 2.68 | 0.09* | 1.24 | 0.97 | 0.89 | 0.52 | 0.10* |
| Klaipėda | 4.61 | 0.75 | 1.93 | −0.97 | 1.26 | 0.83 | 0.97 | 0.47 | −0.21 |
| Laukuva | 5.07 | 0.91 | 2.39 | −0.23 | 1.30 | 0.87 | 0.78 | 0.44 | −0.03* |
| Nida | 4.02 | 1.12 | 2.16 | −0.06* | 1.74 | 1.05 | 0.69 | 0.31 | −0.16* |
| Panevėžys | 5.78 | 0.94 | 3.47 | 0.20* | 1.46 | 1.14 | 0.77 | 0.47 | 0.02* |
| Raseiniai | 5.30 | 0.85 | 2.78 | 0.04* | 1.46 | 1.02 | 0.77 | 0.49 | 0.05* |
| Šiauliai | 5.50 | 1.10 | 2.88 | 0.05* | 1.41 | 1.10 | 0.70 | 0.43 | −0.08* |
| Šilutė | 5.19 | 0.82 | 2.42 | −0.42 | 1.33 | 0.98 | 0.97 | 0.48 | 0.01* |
| Telšiai | 4.88 | 0.91 | 2.14 | −0.37 | 1.20 | 0.86 | 0.76 | 0.47 | −0.06* |
| Ukmergė | 6.25 | 1.28 | 3.37 | 0.42 | 1.35 | 1.29 | 0.84 | 0.53 | 0.03* |
| Utena | 6.34 | 1.26 | 3.78 | 0.39 | 1.43 | 1.23 | 0.85 | 0.46 | 0.05* |
| Varėna | 6.88 | 0.94 | 3.89 | 0.46 | 1.23 | 1.17 | 0.99 | 0.52 | 0.24* |
| Vilnius | 5.76 | 0.95 | 2.97 | 0.03* | 1.25 | 0.87 | 0.84 | 0.49 | 0.21* |

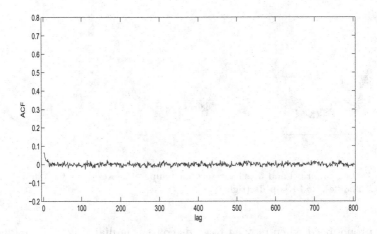

Fig. 3.7    The ACF of the final residuals for Vilnius.

($p < 0.01$ for all stations), clearly rejecting the hypothesis of normality. However, the descriptive statistics (not presented here) show that the final residuals are close to normally distributed. Moreover, with this extent of data it is very difficult to achieve formal proof of normality, since the KS test is sensitive to even small deviations from normality. Such deviations

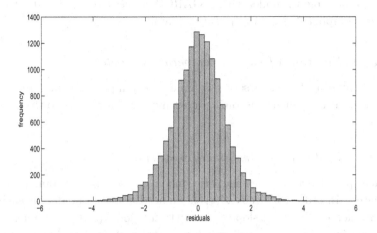

Fig. 3.8   Histogram of the final residuals for Vilnius.

just accumulate for large samples. Thus the Gaussian assumption seems to be reasonable.

The ACF for the squared residuals (see Fig. 3.9) decays rapidly for the first few lags and then varies around zero, indicating that we managed to remove the seasonality in the variance. As it was shown in [Šaltytė Benth

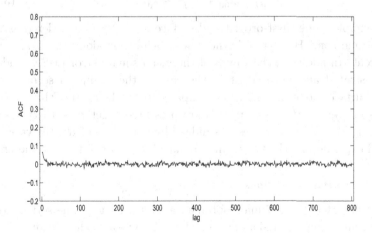

Fig. 3.9   The ACF of the squares of the final residuals for Vilnius.

and Benth (2012)], the significant autocorrelations for the several first lags might be possible to model with a GARCH process incorporated into the residual component in a multiplicative manner.

### 3.2.4    *Estimation of spatial temperature model*

After estimating the time series dynamics of temperature marginally at each given measurement station, we next move our attention to the spatial model.

#### 3.2.4.1    *Spatial model for temporal parameters*

At each spatial location the time series model containing 16 parameters was estimated. In order to specify $Z(\mathbf{s};t)$, we need a spatial model for each of the parameters. Since the slope $a_1$ in the trend model (3.10) is varying little from location to location, we assume it to be constant (equal to the mean of $9.46 \times 10^{-5}$) over the space. Thus we deal with 15 parameters at each of the 16 stations. We consider the trend-surface model (3.12), and start with fitting the second-order model to each set of parameters separately. We observe that the behaviour along $y$ axis is quite stable over the considered spatial area for all parameters. Moreover, the parameters for the cross-term $xy$ were more or less zero in all cases. We therefore choose to work with the simplified second-order trend-surface model given by

$$P^\eta(x, y) = \lambda_{00}^\eta + \lambda_{10}^\eta x + \lambda_{01}^\eta y + \lambda_{20}^\eta x^2. \tag{3.13}$$

A simple plane (first-order trend-surface model) was considered as well for comparison. However, the nonlinear behaviour along the $x$ axis was apparent. In addition, the values of the mean square error (MSE) and $R^2$ demonstrated an improved fit in the case of the simplified second-order trend-surface model in (3.13) as compared to a plane. In Table 3.3, the values of trend-surface parameters are presented together with the corresponding MSE and $R^2$ values. As judged by the values of $R^2$, the variation is well explained by the trend-surface model for most of the parameters.

#### 3.2.4.2    *Spatial correlations*

We argued that the residuals obtained after fitting the time series model were normally distributed and uncorrelated in time at each location. Therefore, we can explore their spatial properties by simply calculating the empirical correlations between all pairs of stations. A scatter plot of 120 values of empirical correlations against the distances between the corresponding

Table 3.3 Fitted parameters of the simplified first-order trend-surface with the corresponding mean square error (MSE) and $R^2$ for temperature.

| Parameter | $\lambda_{00}$ | $\lambda_{10}$ | $\lambda_{01} \times 10^{-10}$ | $\lambda_{20}$ | MSE | $R^2$ (%) |
|---|---|---|---|---|---|---|
| $a_0$ | 6.55 | −0.03 | −0.98 | −0.10 | 0.095 | 65.0 |
| $a_2$ | 6.76 | 0.94 | 0.75 | −0.19 | 0.063 | 86.1 |
| $a_3$ | 7.76 | −0.04 | −3.97 | 0.04 | 0.023 | 10.5 |
| $\alpha_1$ | 0.93 | −0.02 | 3.01 | −0.01 | $2.5 \times 10^{-4}$ | 38.5 |
| $\alpha_2$ | −0.21 | −0.02 | −2.41 | 0.01 | $2.3 \times 10^{-4}$ | 30.0 |
| $\alpha_3$ | 0.10 | −0.002 | −59.2 | −0.001 | $4.2 \times 10^{-5}$ | 37.7 |
| $b_1$ | 4.64 | 0.69 | 0.29 | −0.004 | 0.149 | 75.0 |
| $b_2$ | 0.88 | −0.01 | 5.32 | 0.05 | 0.020 | 28.6 |
| $b_3$ | 2.14 | 0.68 | 0.18 | −0.01 | 0.077 | 84.0 |
| $b_4$ | −0.44 | 0.61 | 9.19 | −0.13 | 0.051 | 70.6 |
| $b_5$ | 1.42 | 0.04 | −0.12 | −0.01 | 0.025 | 6.8 |
| $b_6$ | 0.95 | 0.12 | −8.36 | 0.01 | 0.014 | 50.7 |
| $b_7$ | 0.87 | −0.13 | −8.46 | 0.06 | 0.010 | 18.7 |
| $b_8$ | 0.42 | 0.09 | 3.62 | −0.03 | 0.003 | 31.9 |
| $b_9$ | −0.12 | 0.17 | 8.63 | −0.05 | 0.011 | 40.3 |

stations is presented in Fig. 3.10. Distances were calculated in meters using the program ArcView GIS 3.2a, and scaled by $10^{-5}$ for analyses. Two anisotropic correlation functions were fitted to the empirical correlations: the exponential correlation function defined as

$$\rho_{\exp}(\mathbf{h_s}, \theta_{\mathbf{s}}^{\exp}) = \exp\left(-\frac{1}{\theta_1^{\exp}}\sqrt{\theta_2^{\exp}h_1^2 + h_2^2}\right), \qquad (3.14)$$

and the spherical correlation function

$$\rho_{\mathrm{sph}}(\mathbf{h_s}, \theta_{\mathbf{s}}^{\mathrm{sph}}) = 1 - \frac{3}{2}\frac{\sqrt{\theta_2^{\mathrm{sph}}h_1^2 + h_2^2}}{\theta_1^{\mathrm{sph}}} + \frac{1}{2}\left(\frac{\sqrt{\theta_2^{\mathrm{sph}}h_1^2 + h_2^2}}{\theta_1^{\mathrm{sph}}}\right)^3, \qquad (3.15)$$

for $\mathbf{h_s} = (h_1, h_2)^T$. Their isotropic versions are obtained when the anisotropy parameters $\theta_2^{\exp}$ and $\theta_2^{\mathrm{sph}}$ are equal to one (see [Cressie (1993)] for more details).

The best fit, as judged by the smallest MSE, was obtained by using the anisotropic spherical correlation function. This indicates that the temperature varies differently in different directions in the considered region. The parameter estimates were $\widehat{\theta}_1^{\mathrm{sph}} = 11.92$ and $\widehat{\theta}_2^{\mathrm{sph}} = 1.38$ (both significant at 5% level) with MSE=0.0103. In Fig. 3.10 (left), the fitted function is plotted together with the empirical correlations. The behaviour of the spatial dependencies over the region is close to linear, as seen from Fig. 3.10

(right), where the second best fitting spherical isotropic correlation function is plotted together with the empirical correlations (parameter estimate for this isotropic function is $\widehat{\theta}_1^{\mathrm{sph}} = 1.1 \times 10^3$ with MSE=0.0104).

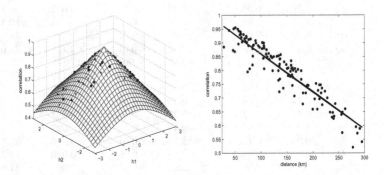

Fig. 3.10   The empirical and fitted spatial correlation functions of temperature (the anisotropic (left) and isotropic (right) spherical correlation function).

### 3.2.4.3   *Model validation for temperature*

The two stations, Palanga and Trakų Vokė, were not used in estimating the spatial-temporal model for temperature in Lithuania. These will be used for model validation. In Palanga, the time series of temperature ranges from 1 June 1992 to 31 August 2004, resulting in 4,475 observations. The time series in Trakų Vokė are much longer, ranging from 1 June 1971 to 31 August 2004 giving us 12,146 observations for model validation. The choice of these two stations is based on their geographical coordinates. Palanga is outside the area made up by 16 stations, but close to the stations on the coast. Trakų Vokė is, on the other hand, inside the study area and located close to Vilnius station (just 12 km away).

In Table 3.4 the values of the parameters estimated for the time series at Palanga and Trakų Vokė stations, respectively, are presented together with their standard errors. In addition, the values of parameters obtained from the fitted spatial model are given. The model fits temperature quite well. Most of the estimated values for Palanga are within two standard errors from the fitted values. Deviations between fitted and estimated values are a bit larger for Trakų Vokė.

Table 3.4 Parameter estimates (Estimated) at Palanga and Trakų Vokė stations with the corresponding values of the standard errors (SE) and values obtained from the spatial model (Fitted).

| Parameter | Palanga | | | Trakų Vokė | | |
|---|---|---|---|---|---|---|
| | Estimated | SE | Fitted | Estimated | SE | Fitted |
| $a_0$ | 7.09 | 0.224 | 6.55 | 5.90 | 0.153 | 5.80 |
| $a_2$ | 6.29 | 0.163 | 6.84 | 7.76 | 0.107 | 7.91 |
| $a_3$ | 7.59 | 0.158 | 7.76 | 8.24 | 0.107 | 7.95 |
| $\alpha_1$ | 0.91 | 0.015 | 0.93 | 0.97 | 0.009 | 0.91 |
| $\alpha_2$ | −0.20 | 0.020 | −0.21 | −0.25 | 0.012 | −0.19 |
| $\alpha_3$ | 0.10 | 0.015 | 0.10 | 0.09 | 0.009 | 0.09 |
| $b_1$ | 4.89 | 0.301 | 4.70 | 5.47 | 0.204 | 6.41 |
| $b_2$ | 0.65 | 0.423 | 0.88 | 0.73 | 0.286 | 1.18 |
| $b_3$ | 2.09 | 0.418 | 2.20 | 2.70 | 0.286 | 3.83 |
| $b_4$ | −1.28 | 0.423 | −0.39 | −0.29 | 0.286 | 0.28 |
| $b_5$ | 1.30 | 0.423 | 1.42 | 1.28 | 0.281 | 1.43 |
| $b_6$ | 0.62 | 0.418 | 0.96 | 0.66 | 0.286 | 1.29 |
| $b_7$ | 1.17 | 0.423 | 0.86 | 0.95 | 0.281 | 0.98 |
| $b_8$ | 0.59 | 0.423 | 0.43 | 0.32 | 0.286 | 0.43 |
| $b_9$ | −0.10 | 0.423 | −0.11 | 0.31 | 0.286 | 0.01 |

### 3.2.5 *A critical view on temporal temperature modelling*

The model proposed and estimated in the subsections above shares some similarities with the time series dynamics proposed for temperature by [Campbell and Diebold (2005)]. The model by [Campbell and Diebold (2005)] has gained some popularity in the field of weather derivatives. Both our model and the one by [Campbell and Diebold (2005)] are similar to or nest a number of related models, for example, the models by [Dornier and Querel (2000)], [Alaton, Djehiche and Stillberger (2002)] and [Cao and Wei (2004)]. The comparison of various models for the time dynamics of temperature, in terms of their performance when forecasting weather indices, has been done by many authors (see, for example, [Oetoma and Stevenson (2005)], [Svec and Stevenson (2007)], [Papazian and Skiadopoulos (2010)], [Zapranis and Alexandridis (2008)] and [Schiller, Seidler and Wimmer (2010)]). The models of [Benth and Šaltytė-Benth (2007)] and [Campbell and Diebold (2005)] are both based on seasonal AR processes, however with important differences in structure influencing their applicability in the field of temperature derivatives. Our goal here is to point out the principle differences between the two models and discuss some important issues to be addressed when modelling the temperature dynamics with the application in weather derivatives markets in mind. The following discussion is extracted from [Šaltytė Benth and Benth (2012)].

It has been demonstrated that the temperature dynamics at a single spatial location follows an AR($p$) process. We therefore skip the MA($q$) part in (3.9) and omit the subindex $k$, denoting spatial location. To make our time series model comparable to the model of [Campbell and Diebold (2005)], we rewrite (3.8) in the following way:

$$Z(t) - \Lambda(t) = \sum_{i=1}^{p} \alpha_i(Z(t-i) - \Lambda(t-i)) + \varepsilon(t). \qquad (3.16)$$

By this we underline that the *deseasonalized* temperature follows an AR($p$) process. In other words, today's deseasonalized temperatures are regressed on deseasonalized temperatures at earlier time points when estimating the AR structure of the temperature evolution. Thus, in our time series dynamics, the seasonality is modelled explicitly as the level towards which the temperatures are mean-reverting. By assuming that the residual process $\varepsilon(t)$ has mean zero, the expected temperature follows the recursion

$$\mathbb{E}[Z(t)] - \Lambda(t) = \sum_{i=1}^{p} \alpha_i(\mathbb{E}[Z(t-i)] - \Lambda(t-i)). \qquad (3.17)$$

Assuming in addition stationarity of the coefficients $\alpha_i$ of the AR($p$) process, $\Lambda(t)$ becomes the stationary mean of $Z(t)$. In this way, it is natural to interpret $\Lambda(t)$ as the *temperature seasonality*, a crucial model component in financial applications. The seasonal mean of temperature is the main factor explaining, for instance, the temperature indices CAT, HDD, CDD (see [Šaltytė Benth and Benth (2012)] for a detailed empirical analysis of this).

[Dornier and Querel (2000)] though represent the mean process $\mu(t)$ in a slightly different way

$$\mu(t) = \Lambda_1(t) + \sum_{i=1}^{p} \alpha_i Z(t-i), \qquad (3.18)$$

for a deterministic function $\Lambda_1(t)$. This function does not become the stationary mean of temperature, but rather a deterministic component in the AR dynamics. [Campbell and Diebold (2005)] call this function *seasonal component* or *seasonality* of the temperature dynamics, which is not the same as *seasonal mean function*. In other words, [Campbell and Diebold (2005)] model seasonality of the temperature indirectly, through the summation of the seasonal component $\Lambda_1(t)$ and the AR structure. Furthermore, they regress today's *deseasonalized* temperature on the previous days' temperature (not deseasonalized). This contrast is very important for both the interpretation and application of the model.

As indicated, to reveal the seasonality structure in the model of [Campbell and Diebold (2005)] some mathematical computations using the AR structure are required. In [Šaltytė Benth and Benth (2012)], it is shown how the true seasonal function could be derived if the seasonal component in the model of [Campbell and Diebold (2005)] was known. All parameters of the seasonal function $\Lambda(t)$ are then functions of AR parameters, $\alpha_i$, implying that it is necessary to know AR parameters in order to find the true seasonality function. Obviously, this increases uncertainty of the parameter estimates, and thus the uncertainty in the identification of the mean seasonal temperature structure. This is a problem in view of the apparent need to have an accurate estimate of the temperature seasonality function $\Lambda(t)$ in applications to weather markets. Calculating the seasonality function from the seasonal component may result in a wrong specification, including an excessive number of parameters and uncertainty in the estimates.

Our temperature dynamics model (introduced in [Benth and Šaltytė-Benth (2007)] and further generalized in [Šaltytė Benth and Benth (2012)]) differs from the one by [Campbell and Diebold (2005)] by its simplicity in terms of number of parameters. Even though our model contains far less parameters, it is still sufficiently sophisticated in order to explain the basic stylized facts of temperature as good as the parameter-intensive alternative proposed by [Campbell and Diebold (2005)]. We promote the decomposition approach when estimating our model, which turns out to be advantageous when the aim is to build a confident time series dynamics. This approach leads to a very low-order AR process in contrast to [Campbell and Diebold (2005)], suggesting in their analysis to use 25 lags in the AR dynamics.

We do not consider the GARCH dynamics in modelling the seasonal heteroskedastic residuals in this book (although we discuss a stochastic volatility model in Chapter 6). However, it has been shown by the authors (see [Šaltytė Benth and Benth (2012)]), that a *multiplicative* GARCH component in addition to the seasonal deterministic volatility function might be beneficial in improving the model fit. [Campbell and Diebold (2005)] argue for an *additive* GARCH component, which potentially can lead to negative variance in addition to making the estimation of the model more complicated than necessary.

For a detailed discussion of these important issues, the reader is referred to the paper by [Šaltytė Benth and Benth (2012)].

### 3.2.6　*Estimation of wind speed model*

The model for wind speed was estimated on in-sample data from the period 1 January 1977 to 31 December 1979, consisting of 10,950 observations recorded in 16 stations. The additional observation series were left out for model validation. Since the distribution of wind speed data was clearly skewed (see Fig. 2.6), we used the logarithmic transformation at each spatial location separately. The logarithmically transformed data exhibit a more symmetric pattern (Fig. 3.11), which makes modelling simpler. Motivated

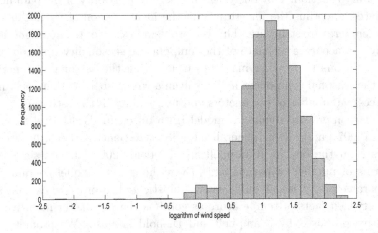

Fig. 3.11　The histogram for logarithmically transformed wind speed observations for Vilnius.

by the study in [Šaltytė Benth and Benth (2010)] for wind speed in New York, we also checked out the square root and other power transformations for the same purpose, however, they did not perform better than the logarithmic transformation for the Lithuanian wind speed data. We recall from [Šaltytė Benth and Benth (2010)] that for the case of New York, the Box-Cox power transformation gave more symmetric and normally distributed data than the logarithmic one.

The simple descriptive statistics of logarithmically transformed wind speed data are presented in Table 3.5. Even though the logarithmic transformation performed well, skewness and kurtosis still remain quite different from zero. In addition, the ACF of logarithmically transformed wind speeds (Fig. 3.12) demonstrates rather clear seasonal effects in the data, which need to be modelled.

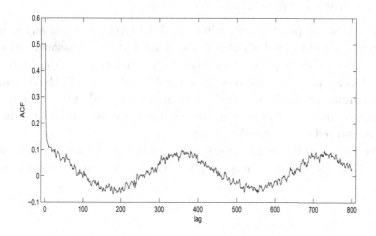

Fig. 3.12   The ACF for logarithmically transformed wind speed observations for Vilnius.

The estimation procedure for the wind speed model is analogous to the one for temperature. We therefore discuss here only briefly the most important points and differences. A detailed description of the estimation procedure can be found in [Šaltytė Benth and Šaltytė (2011)].

Table 3.5   Descriptive statistics for the logarithmically transformed wind speed records ('Std' stands for standard deviation).

| Station | Mean | Std | Min | Max | Skewness | Kurtosis |
|---|---|---|---|---|---|---|
| Biržai | 1.14 | 0.45 | −2.30 | 2.48 | −0.48 | 0.61 |
| Dotnuva | 0.86 | 0.52 | −2.30 | 2.67 | 0.003 | −0.18 |
| Kaunas | 1.25 | 0.46 | −1.20 | 2.57 | −0.43 | 0.32 |
| Kybartai | 1.03 | 0.54 | −2.30 | 2.94 | −0.36 | 0.70 |
| Klaipėda | 1.40 | 0.54 | −0.69 | 3.13 | −0.15 | −0.36 |
| Laukuva | 1.12 | 0.55 | −2.30 | 2.93 | −0.11 | 0.47 |
| Nida | 1.38 | 0.53 | −2.30 | 2.92 | −0.44 | 0.81 |
| Panevėžys | 1.09 | 0.48 | −2.30 | 2.40 | −0.70 | 1.21 |
| Raseiniai | 1.25 | 0.41 | −2.30 | 2.59 | −0.43 | 1.09 |
| Šiauliai | 0.94 | 0.48 | −2.30 | 2.23 | −0.99 | 2.74 |
| Šilutė | 1.22 | 0.50 | −2.30 | 2.79 | −0.42 | 1.03 |
| Telšiai | 1.04 | 0.42 | −2.30 | 2.53 | −0.49 | 1.16 |
| Ukmergė | 1.16 | 0.54 | −2.30 | 2.77 | −0.65 | 1.56 |
| Utena | 0.81 | 0.55 | −2.30 | 2.20 | −0.78 | 1.52 |
| Varėna | 0.72 | 0.56 | −2.30 | 2.14 | −0.83 | 1.54 |
| Vilnius | 1.13 | 0.44 | −2.30 | 2.36 | −0.43 | 0.81 |

### 3.2.6.1    *Seasonal component and ARMA process*

In contrast to temperature, the wind speed data did not reveal any significant linear trend throughout the period of interest. The seasonal function
(3.10) with only three parameters was sufficient to remove seasonal fluctuations in wind speed observations. In Table 3.6 (columns 1-3), the estimates
of the parameters of the seasonal function are reported. Even though the
seasonal component is present in the wind speed observations, it is much
less pronounced as compared to temperature.

The ACF and PACF for deseasonalized wind speed data are presented
in Fig. 3.13 and Fig. 3.14, respectively.   Clearly, seasonal variations were

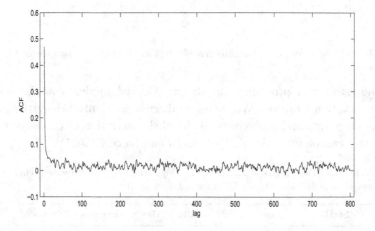

Fig. 3.13    The ACF for deseasonalized wind speed observations for Vilnius.

removed from data, however, strong autocorrelations are present in the
residuals.  As indicated by the PACF plot, we need a higher order AR
process to explain the autocorrelations.  To determine the order of the
AR process, we used the AIC, the Box-Ljung test for ACF and assessed
histograms of residuals. The order of AR process was bounded to 10 in our
analysis. We started from the lowest order and proceeded with estimating
the higher order AR processes. According to the model selection criteria
specified above, the AR(3) or AR(4) process was sufficient for modelling
the autocorrelations in the residuals in most of stations. However, in four
stations a higher order AR was required to obtain a satisfactory model
fit. Clearly, the wind speed is much more unstructured and noisy variable
than temperature. We remark that an ARMA process with combinations

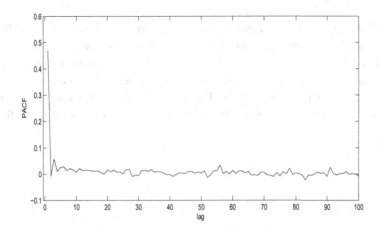

Fig. 3.14   The PACF for deseasonalized wind speed observations for Vilnius.

Table 3.6   Fitted parameters of the seasonal component and AR process. Asterisk marks parameters not significantly different from zero at the level of 5%.

| Stations | $a_0^k$ | $a_2^k$ | $a_3^k$ | $\alpha_1^k$ | $\alpha_2^k$ | $\alpha_3^k$ | $\alpha_4^k$ | $\alpha_5^k$ | $\alpha_6^k$ | $\alpha_7^k$ | $\alpha_8^k$ |
|---|---|---|---|---|---|---|---|---|---|---|---|
| Biržai | 1.14 | 0.15 | 0.04 | 0.46 | −0.05 | 0.05 | 0.02 | | | | |
| Dotnuva | 0.86 | 0.14 | 0.08 | 0.49 | −0.04 | 0.08 | 0.09 | 0.03 | 0.06 | | |
| Kaunas | 1.25 | 0.20 | 0.03 | 0.47 | −0.05 | 0.05 | | | | | |
| Kybartai | 1.03 | 0.19 | 0.03 | 0.48 | −0.02* | 0.06 | | | | | |
| Klaipėda | 1.40 | 0.18 | −0.06 | 0.52 | −0.05 | 0.06 | 0.03 | | | | |
| Laukuva | 1.11 | 0.17 | 0.03 | 0.48 | −0.04 | 0.05 | 0.04 | | | | |
| Nida | 1.39 | 0.18 | −0.05 | 0.47 | 0.002* | 0.07 | 0.04 | 0.04 | 0.03 | 0.05 | 0.07 |
| Panevėžys | 1.09 | 0.14 | 0.05 | 0.48 | −0.03 | 0.05 | 0.04 | | | | |
| Raseiniai | 1.25 | 0.14 | 0.06 | 0.43 | −0.04 | 0.05 | 0.02 | | | | |
| Šiauliai | 0.94 | 0.16 | 0.06 | 0.47 | −0.03 | 0.05 | 0.02 | | | | |
| Šilutė | 1.22 | 0.11 | 0.06 | 0.47 | −0.04 | 0.07 | | | | | |
| Telšiai | 1.03 | 0.12 | 0.03 | 0.48 | −0.05 | 0.06 | | | | | |
| Ukmergė | 1.16 | 0.17 | 0.04 | 0.46 | −0.04 | 0.07 | 0.02 | | | | |
| Utena | 0.82 | 0.18 | 0.06 | 0.45 | −0.01* | 0.06 | 0.02 | 0.03 | 0.05 | | |
| Varėna | 0.73 | 0.15 | 0.07 | 0.47 | −0.03 | 0.04 | 0.03 | −0.00* | 0.05 | | |
| Vilnius | 1.13 | 0.16 | 0.03 | 0.47 | −0.03 | 0.06 | | | | | |

of AR and MA components neither did improve the model fit nor resulted in a model with fewer parameters. The estimates of the AR parameters are reported in Table 3.6 (starting in column 4). It is remarkable that the parameters of the AR process are very homogeneous over space.

### 3.2.6.2   *Residuals*

A positive kurtosis and negative skewness were observed in the residuals obtained after eliminating the AR process from the deseasonalized wind speed observations at all stations. However, the distributions were reasonably symmetric as judged by histograms (not shown). Although the residuals are not autocorrelated in time (Fig. 3.15), the ACF for squared residuals reveals clearly the presence of time-dependent variations (see Fig. 3.16).

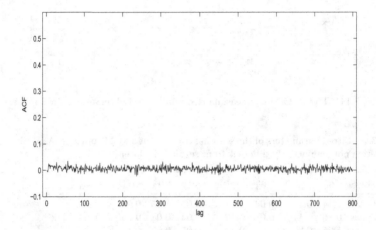

Fig. 3.15   The ACF for residuals obtained after eliminating seasonal effects and AR process from the wind speed observations for Vilnius.

The analysis of squared residuals was performed in the same way as for temperature. The seasonal variance clearly seen in Klaipėda (Fig. 3.17, left) is common for most stations in Lithuania. A less pronounced seasonal behaviour as in Vilnius (Fig. 3.17, right) was observed in a few other stations. As we demonstrated earlier, the seasonal pattern is more complex and pronounced for temperature, but there is a clear resemblance between the two. For both wind speed and temperature, the ACF of squared residuals exhibits a clear seasonal pattern, while the residuals themselves are uncorrelated in time.

Since the pattern of the time-dependent variance seems to be rather flat with lots of variations (see Fig. 3.17), we restrict our attention to the function (3.11) with $L_\sigma = 1$. In addition, we use two more variance functions, the daily empirical and time-independent (constant) empirical variance, for modelling the heteroskedasticity in the residuals. The parameter estimates

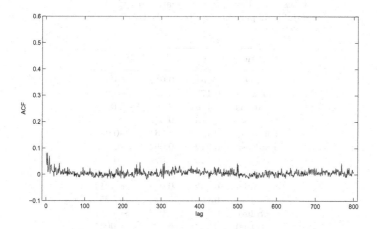

Fig. 3.16   The ACF for squared residuals obtained after eliminating seasonal effects and AR process from the wind speed observations for Vilnius.

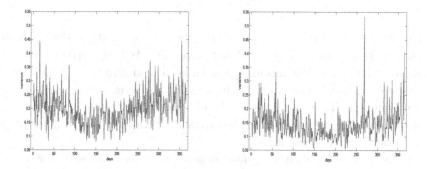

Fig. 3.17   The empirical variance for wind speed residuals for Klaipėda (left) and Vilnius (right).

for $\sigma_k^2(t)$ are given in Table 3.7 for each station; here the values in the first column correspond to the empirical variance in the particular station.

The residuals were next normalized by dividing them by the square root of one of the three considered variance functions. The residuals were closest to the white-noise when the constant empirical variance of the residuals was used. This is likely a consequence of a high noise level in data, where the seasonal pattern is out-competed by noise. The ACF for the residuals and squared residuals normalized by the empirical variance are presented in

Table 3.7   Parameter estimates for $\sigma_k^2(t)$; insignificant parameters are marked with an asterisk.

| Station | $b_1^k$ | $b_2^k$ | $b_3^k$ |
|---|---|---|---|
| Biržai | 0.151 | 0.012 | −0.004* |
| Dotnuva | 0.177 | 0.030 | −0.004 |
| Kaunas | 0.148 | 0.011 | −0.003* |
| Kybartai | 0.213 | 0.048 | −0.002* |
| Klaipėda | 0.198 | 0.035 | −0.017 |
| Laukuva | 0.219 | 0.064 | −0.029 |
| Nida | 0.169 | 0.022 | −0.018 |
| Panevėžys | 0.165 | 0.007* | −0.013 |
| Raseiniai | 0.132 | 0.021 | −0.001* |
| Šiauliai | 0.166 | 0.014 | −0.005* |
| Šilutė | 0.189 | 0.052 | −0.027 |
| Telšiai | 0.133 | 0.023 | −0.005* |
| Ukmergė | 0.221 | 0.014 | −0.017 |
| Utena | 0.210 | 0.027 | −0.009* |
| Varėna | 0.229 | 0.043 | −0.019 |
| Vilnius | 0.143 | 0.030 | −0.001* |

Fig. 3.18 (left and right, respectively). According to the ACF, the residuals seem to be a zero-mean uncorrelated noise. The ACF of squared residuals decays for the first few lags and then varies around zero, clearly indicating that we managed to remove the heteroskedasticity in the residuals. The behaviour of the ACF for residuals and squared residuals in the case of fitted seasonal and empirical daily variance functions is nearly identical to the described case (not shown).

The final residuals have a small negative skewness. A negative kurtosis is pointing towards a less peaky distribution than the normal. In spite of this, the distribution seems to be close to normal (see Fig. 3.19) with mean and standard deviation basically equal zero and one, respectively.

To justify a choice of the constant empirical variance in modelling the time dependency in residuals, we simulated 100 paths, each consisting of 365 values, for each of the three considered variance functions. The average of the obtained mean square simulation errors (MSSE) was calculated at each station. For most of stations, the MSSE was smallest in the case of the constant empirical variance function. In five stations the fitted seasonal variance performed slightly better than the constant empirical variance, while there were two stations where daily empirical variance outperformed the constant empirical variance. However, MSSE values were very close

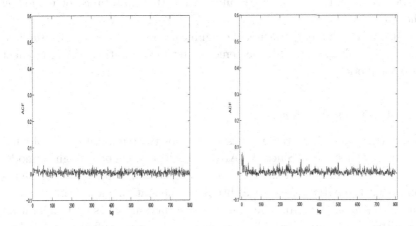

Fig. 3.18 The ACF for residuals (left) and squared residuals (right) obtained after eliminating empirical variance in wind speed residuals for Vilnius.

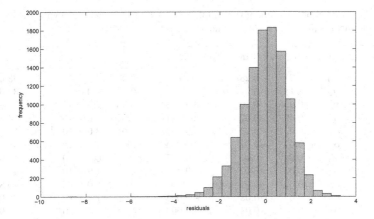

Fig. 3.19 The histogram for residuals obtained after eliminating seasonal variance in wind speed residuals for Vilnius.

to each other for the considered cases. Therefore, taking into account a number of parameters to be estimated, the choice of the constant emiprical variance is supported. A similar simulation study was performed under the assumption that the residuals obtained after eliminating the AR process from the deseasonalized wind speed observations was white noise. The

MSSE became 1.5 times larger, illustrating the importance of modelling the variance.

Just as for the temperature, the significant autocorrelations for squared residuals could presumably be removed by incorporating GARCH effect into the model.

### 3.2.6.3  *Spatial modelling*

There were up to 12 parameters estimated for the temporal model of wind speed at each station. Since the AR(3) or AR(4) was not enough to model the variations in deseasonalized wind speed data in some stations, we assume that the AR(8) process was fitted at each station and set unestimated parameters equal to zero. The same trend-surface model as for temperature was fitted for wind speed. The parameters of the simplified trend-surface model are reported in Table 3.8. As indicated by the small MSE and reasonably high values of $R^2$, the variation is explained quite well in most of the spatial parameters by the second-order trend-surface model.

Table 3.8   Fitted parameters of the simplified first-order trend-surface with the corresponding mean square error (MSE) and $R^2$ for wind speed.

| Parameter | $\lambda_{00}$ | $\lambda_{10}$ | $\lambda_{01} \times 10^{-10}$ | $\lambda_{20}$ | MSE | $R^2$ (%) |
|---|---|---|---|---|---|---|
| $a_0$ | 1.31 | −0.28 | 0.03 | 0.05 | 0.028 | 63.0 |
| $a_2$ | 0.18 | −0.02 | −0.01 | 0.01 | $6.9 \times 10^{-4}$ | 34.6 |
| $a_3$ | −0.03 | 0.10 | 0.0003 | −0.03 | $7.4 \times 10^{-4}$ | 77.8 |
| $\alpha_1$ | 0.49 | −0.02 | −0.0001 | 0.002 | $3.6 \times 10^{-4}$ | 44.8 |
| $\alpha_2$ | −0.01 | −0.03 | −0.008 | 0.01 | $3.0 \times 10^{-4}$ | 52.4 |
| $\alpha_3$ | 0.07 | −0.01 | 0.0003 | 0.004 | $1.2 \times 10^{-4}$ | 38.5 |
| $\alpha_4$ | 0.01 | 0.01 | 0.007 | −0.002 | $6.5 \times 10^{-4}$ | 19.6 |
| $\alpha_5$ | 0.02 | −0.02 | 0.001 | 0.01 | $1.9 \times 10^{-4}$ | 44.2 |
| $\alpha_6$ | 0.02 | −0.01 | −0.01 | 0.01 | $5.0 \times 10^{-4}$ | 40.8 |
| $\alpha_7$ | 0.02 | −0.03 | −0.003 | 0.01 | $1.3 \times 10^{-5}$ | 59.9 |
| $\alpha_8$ | 0.03 | −0.04 | −0.003 | 0.01 | $2.5 \times 10^{-4}$ | 59.9 |
| $\sigma^2$ | 0.22 | −0.03 | −0.02 | 0.04 | 0.001 | 36.0 |

The empirical correlations calculated between all 120 pairs of stations are plotted against the corresponding distances (in kilometers) in Fig. 3.20. We considered four types of spatial correlation functions: anisotropic spherical and exponential, and their isotropic alternatives. The isotropic and anisotropic exponential functions fitted data best according to the MSE. However, the anisotropy parameter was close to one and MSE was slightly smaller for the isotropic exponential function

$$\rho_{exp}(\mathbf{h_s}, \theta^{\exp}) = \exp\left(-\frac{1}{\theta^{\exp}}h\right),\qquad(3.19)$$

where $h$ is a distance between pairs of stations. The parameter of the exponential spatial correlation function was estimated to be $\theta^{\exp} = 344.8$ with the corresponding MSE equal to 0.052. The parameter of anisotropy being almost equal to one demonstrates that the behaviour of wind speed over the considered region is more or less the same in different directions. The fitted exponential function is presented in Fig. 3.20.

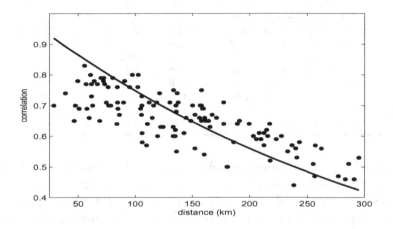

Fig. 3.20  The empirical spatial correlations and fitted isotropic exponential function.

### 3.2.6.4  *Model validation for wind speed*

We validate the wind speed model in a different way than the model for temperature. The purpose is to illustrate the various approaches that can be used for model validation. We first validate the temporal model on out-of-sample observations. Then the validation of the spatial model is performed using two stations which were excluded from the estimation procedure.

For the validation of the temporal model we use 365 available out-of-sample wind speed observations from the year 2007 for all 16 stations. To validate the temporal model, we generate one-step-ahead predictions for out-of-sample observations and calculate the prediction errors (PE) defined as the differences between the observations and predictions. Simple statistical tests are then performed on PEs.

Table 3.9   Descriptives (mean and standard deviation (Std)), KS test p-value for prediction errors (KS p-value), and percentage of observations outside 95% prediction intervals for empirical (EMP), daily empirical (DEMP) and fitted variance (FIT) functions.

| Station | mean | Std | KS p-value | EMP | DEMP | FIT |
|---|---|---|---|---|---|---|
| Biržai | −0.03 | 0.34 | 0.243 | 1.68 | 2.24 | 1.96 |
| Dotnuva | −0.01 | 0.34 | 0.701 | 0.28 | 1.96 | 1.12 |
| Kaunas | −0.01 | 0.35 | 0.350 | 2.52 | 3.64 | 2.52 |
| Kybartai | 0.01 | 0.41 | 0.105 | 0.56 | 1.96 | 1.12 |
| Klaipėda | −0.02 | 0.38 | 0.617 | 1.96 | 2.52 | 2.24 |
| Laukuva | 0.04 | 0.40 | 0.471 | 2.24 | 2.80 | 2.80 |
| Nida | −0.05 | 0.38 | 0.481 | 4.20 | 5.32 | 3.92 |
| Panevėžys | −0.01 | 0.36 | 0.087 | 3.08 | 5.32 | 3.36 |
| Raseiniai | −0.03 | 0.34 | 0.218 | 4.48 | 5.32 | 5.04 |
| Šiauliai | −0.04 | 0.37 | 0.015 | 3.92 | 4.76 | 4.48 |
| Šilutė | 0.06 | 0.38 | 0.863 | 2.52 | 4.48 | 3.64 |
| Telšiai | 0.04 | 0.30 | 0.293 | 1.40 | 3.08 | 1.68 |
| Ukmergė | −0.04 | 0.40 | 0.265 | 2.52 | 3.36 | 2.52 |
| Utena | −0.08 | 0.55 | 0.006 | 8.12 | 10.36 | 8.12 |
| Varėna | −0.08 | 0.45 | 0.064 | 4.76 | 5.60 | 5.32 |
| Vilnius | −0.02 | 0.34 | 0.810 | 2.24 | 3.64 | 1.96 |

The PEs were normally distributed for all but two stations with means and standard deviations given in Table 3.9 (columns 1-2). According to the KS test (corresponding p-values are reported in Table 3.9, column 3), the normality assumption was not met in Šiauliai and Utena. However, the histograms of PEs demonstrated a reasonable symmetry in these stations (not shown).

The prediction intervals (PI) were calculated as follows

$$\hat{Z}_k(t) \pm z_{\alpha/2}\sqrt{\mathrm{Var}(\varepsilon_k(t))}, \tag{3.20}$$

where $\hat{Z}$ denotes the prediction of $Z$ and $z_{\alpha/2}$ is the $\alpha/2$-quantile of a standard normal distribution. An appropriate expression for $\mathrm{Var}(\varepsilon_k(t))$, the variance of the noise component, can be found from the model being applied (see [Chatfield (2000)]). In the case of AR($p$) process, the empirical variance of noise can simply be used. We therefore calculated the PIs for wind speed using constant empirical variance of noise. Moreover, for the purposes of comparison, we also calculated the PIs with the estimated seasonal variance function and daily empirical variance.

In Fig. 3.21, the observed and predicted values for the out-of-sample period with 95% PI are plotted. Only between 0.56% and 8.12% of observations were beyond these bands for wind speed in meteorological stations in Lithuania (see Table 3.9, column 4 for details). As seen from Table 3.9 (columns 5 and 6), the constant empirical variance performs better, or as good as the fitted seasonal variance function for all stations.

It has been shown in several empirical studies that out-of-sample PEs tend to be larger than residuals from model fitting implying too narrow PIs on average (see [Chatfield (2000)] for a discussion). This would mean that often more than 5% of future observations will fall outside of 95% PI on average. In our study we only observe one station with more than 5% of observations beyond the 95% PI. We therefore conclude that the temporal model with empirical variance for residuals fits data well in the region of interest. Notice that the original wind speed data were symmetrized using a logarithmic transformation. The predicted values may easily be back-transformed by using the exponential. The back-transformed PIs can be calculated in an analogous way.

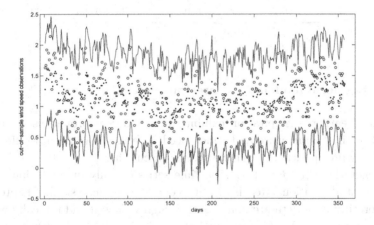

Fig. 3.21 Out-of-sample data and 95% prediction intervals for wind speed for Vilnius. Predictions are marked with points and observations with circles.

Just as for temperature, we use Palanga and Trakų Vokė stations for validation of the spatial model. The estimated parameters for the time series in Palanga and Trakų Vokė are given in Table 3.10 together with the corresponding standard errors and the values of the parameters obtained from the fitted trend-surface model. Most of the estimated values were within or reasonably close to an interval two standard errors from the fitted values, with deviations larger for Trakų Vokė. Moreover, the order of AR process for Trakų Vokė was higher than that for Palanga, indicating higher noise level. We remind that there was no trend in the wind speed model detected. Consequently, the parameter $a_1$ was not estimated.

Table 3.10    Parameters estimated (Estimated) at the out-of-sample stations (Palanga and Trakų Vokė) (T.Vokė) with the corresponding values of the standard errors (SE) and values obtained from the spatial model (Fitted).

| | Palanga | | | T.Vokė | | |
|---|---|---|---|---|---|---|
| $\eta$ | Estimated | SE | Fitted | Estimated | SE | Fitted |
| $a_0$ | 1.41 | 0.006 | 1.26 | 1.31 | 0.005 | 0.96 |
| $a_2$ | 0.14 | 0.009 | 0.15 | 0.18 | 0.007 | 0.17 |
| $a_3$ | 0.01 | 0.009 | −0.03 | 0.03 | 0.007 | 0.48 |
| $\alpha_1$ | 0.48 | 0.014 | 0.49 | 0.43 | 0.010 | 0.47 |
| $\alpha_2$ | −0.05 | 0.015 | −0.03 | −0.02 | 0.010 | −0.03 |
| $\alpha_3$ | 0.07 | 0.015 | 0.06 | 0.04 | 0.010 | 0.06 |
| $\alpha_4$ | 0.06 | 0.014 | 0.02 | | | |
| $\sigma^2$ | 0.17 | 0.005 | 0.19 | 0.22 | 0.005 | 0.20 |

## 3.3    Temporal modelling of precipitation

As noted earlier, precipitation exhibits quite different properties as compared to the temperature and wind speed. We therefore need a different model for precipitation, which we shortly discuss here.

Denote by $P(t)$, $t = 1, ..., T$, a time series of daily precipitation at a given spatial location. Recall from the exploratory analysis in Chapter 2 that precipitation data are non-negative, highly skewed and typically with many zero values. In the case of wind speed with strictly positive values, we considered an AR model for the logarithmically transformed time series. Due to the many zero values however, an initial logarithmic transformation of precipitation is not the right approach. Instead, we assume a multiplicative time series model for the daily precipitation observations.

We denote by $g(t)$ a daily seasonal average amount of precipitation, and let $J(t)$ be the i.i.d. noise component. A product $g(t)J(t)$ describes then the amount of precipitation on a day where precipitation is observed. Denote by $\varepsilon(t)$ an indicator function getting the value of one with probability $\lambda(t)$ (precipitation) and value zero (no precipitation) with probability $1 - \lambda(t)$, where $\lambda(t)$ is the precipitation intensity on day $t$. Then the model for daily precipitation can be defined as

$$P(t) = g(t)J(t)\varepsilon(t).$$

Both, $g(t)$ and $\lambda(t)$ are seasonally dependent functions and are modelled by a truncated Fourier series

$$\delta(t) = c_0^\delta + \sum_{l=1}^{L_\delta} \left( c_{2l}^\delta \cos(2l\pi t/365) + c_{2l+1}^\delta \sin(2l\pi t/365) \right), \qquad (3.21)$$

where $\delta$ is generic notation for $\lambda$ or $g$. We consider a case of deterministic functions here only, though generalizations including ARIMA processes, for example, are possible.

### 3.3.1 *Estimation of precipitation time series model*

Each component in the model for precipitation will be estimated separately. In order to estimate the empirical daily intensity and empirical daily average amount of precipitation for any day of a year, we assume that the time series observations are simply repeated measurements of precipitation. On each of 365 days, we thus have 28 repeated observations in the period January through August and 27 observations in September through December. Only the number of days with observed precipitation during the 28 (or 27) years period are of interest when estimating the daily intensity and daily average amount of precipitation.

The empirical daily intensity for a given day is estimated as the proportion of observed precipitation, that is, the number of years with a precipitation observed on that date divided by the total number of years of observations (correspondingly, 28 or 27). The empirical analysis of precipitation data showed that the function $\lambda(t)$ in (3.21) with $L_\lambda = 2$ describes the intensity at all locations of interest adequately. Parameter estimates for $\lambda(t)$ are presented in Table 3.11 for all stations. Clearly, the daily intensity of precipitation is highest in the winter and lowest in the spring and autumn, with a small bump in the summer (see Fig. 3.22). Note that the shape is very similar to the seasonal volatility of temperatures.

The seasonal daily average amount of precipitation is estimated by finding the empirical average of precipitation on the given day over all the years. The empirical average is based *only* on observations on those days where there was observed precipitation, excluding possible zero precipitation data. The seasonal function (3.21) with $L_g = 2$ fitted data well in all stations, with estimated parameters presented in Table 3.11. As seen from Fig. 3.23,

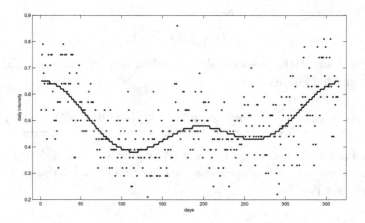

Fig. 3.22   Daily intensity of precipitation for Vilnius.

Table 3.11   Fitted parameters of the intensity function $\lambda(t)$ and daily seasonal average $g(t)$ of precipitation. Parameters significantly not different from zero at the level of 5% are marked with asterisk.

| Stations | $c_0^\lambda$ | $c_1^\lambda$ | $c_2^\lambda$ | $c_3^\lambda$ | $c_4^\lambda$ | $c_0^g$ | $c_1^g$ | $c_2^g$ | $c_3^g$ | $c_4^g$ |
|---|---|---|---|---|---|---|---|---|---|---|
| Biržai | 0.52 | 0.10 | −0.05 | 0.03 | 0.007* | 3.52 | 1.47 | −0.45 | 0.45 | 0.08* |
| Dotnuva | 0.48 | 0.05 | −0.03 | 0.04 | 0.03 | 3.38 | −1.21 | −0.42 | 0.25 | 0.06* |
| Kaunas | 0.50 | 0.07 | −0.02 | 0.04 | 0.02 | 3.52 | −1.32 | −0.50 | 0.29 | 0.12* |
| Kybartai | 0.47 | 0.05 | −0.02 | 0.05 | 0.03 | 3.50 | −1.35 | −0.48 | 0.35 | 0.10* |
| Klaipėda | 0.50 | 0.13 | −0.05 | 0.03 | 0.01* | 4.28 | −0.89 | −1.35 | −0.11* | 0.04* |
| Laukuva | 0.51 | 0.11 | −0.04 | 0.05 | 0.01* | 4.62 | −0.11 | −1.07 | 0.13* | 0.11* |
| Nida | 0.46 | 0.12 | −0.05 | 0.02 | 0.01* | 4.46 | −1.22 | −1.24 | 0.04* | −0.01* |
| Panevėžys | 0.49 | 0.08 | −0.03 | 0.04 | 0.02 | 3.47 | −1.49 | −0.43 | 0.23 | −0.03* |
| Raseiniai | 0.54 | 0.12 | −0.03 | 0.05 | 0.01 | 3.67 | −1.39 | −0.62 | 0.33 | 0.09* |
| Šiauliai | 0.51 | 0.09 | −0.04 | 0.04 | 0.001* | 3.44 | −1.45 | −0.58 | 0.27 | −0.27* |
| Šilutė | 0.47 | 0.11 | −0.05 | 0.03 | 0.02 | 4.83 | −1.54 | −1.49 | 0.07* | 0.16* |
| Telšiai | 0.50 | 0.11 | −0.05 | 0.04 | 0.01* | 4.58 | −1.12 | −0.99 | 0.20* | 0.16* |
| Ukmergė | 0.49 | 0.08 | −0.02 | 0.05 | 0.02 | 3.75 | −1.48 | −0.61 | 0.06* | 0.08* |
| Utena | 0.50 | 0.08 | −0.02 | 0.05 | 0.02 | 3.83 | −1.55 | −0.52 | 0.21* | 0.11* |
| Varėna | 0.48 | 0.07 | −0.001* | 0.06 | 0.02 | 4.09 | −1.58 | −0.63 | 0.15* | 0.14* |
| Vilnius | 0.49 | 0.09 | −0.01 | 0.07 | 0.02 | 3.93 | −1.56 | −0.58 | 0.04* | 0.24 |

the average amount of precipitation is considerably higher in the summer than in the winter, with the variation also highest in the summer.

By dividing the amount of rain on a rainy day, $g(t)J(t)$, by the estimated seasonal daily average function, $g(t)$, we obtain the noise component, $J(t)$. Since we assume $\{J(t)\}_{t=1}^T$ to be i.i.d. draws from a certain distribution,

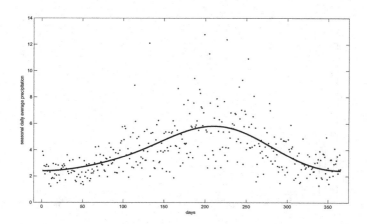

Fig. 3.23  Daily average amount of precipitation for Vilnius.

we can explore the data set consisting only of days with precipitation. The component $J(t)$ is very noisy (see Fig. 3.24) and it is distributed as shown

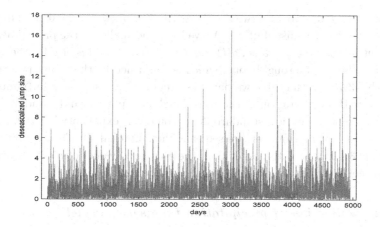

Fig. 3.24  Time series of noise in precipitation on days when precipitation is observed for Vilnius.

in Fig. 3.25. The same pattern is common for all stations. From this, it is reasonable to assume $J(t)$ to be i.i.d., and a natural choice of the distribution of the $J(t)$ component is the exponential one. We recall the

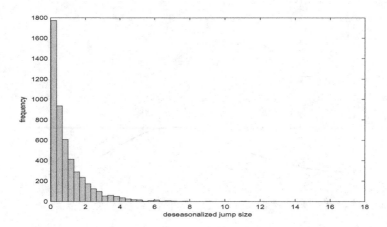

Fig. 3.25   The histogram of noise precipitation on days when precipitation is observed for Vilnius.

exponential distribution to have probability density function

$$f(x; \mu) = \frac{1}{\mu} \exp\left(-\frac{x}{\mu}\right),$$

for $x \geq 0$ and zero otherwise, where the parameter $\mu > 0$ is the expected value of the distribution. As we have "normalized" the precipitation amount by the average value, $g(t)$, we expect the estimates of $\mu$ for the various locations to be roughly one, which is confirmed by the values reported in Table 3.12. In the table we have also reported the sample size (number of days with precipitation in the considered time period) and standard deviations. The theoretical standard deviation of the exponential distribution is also equal to $\mu$. However, we see that the empirical standard deviations in Table 3.12 are slightly larger than the estimated values of $\mu$'s, indicating that other distributions might give an even better fit of $J(t)$.

### 3.3.2    *Validation of precipitation time series model*

We validate the proposed time series model by comparing the observed *monthly* precipitation with the values of precipitation simulated from the model. We simulate 1000 realizations from our model for each day over one year, i.e. 365 days. We first generate two sequences of random innovations, one for simulating the component $\varepsilon(t)$, another for noise component $J(t)$. Then a sequence of values $P(t)$ is computed from the model. We aggregate

Table 3.12 Sample size ($N$), parameter estimates for $\mu$, and standard deviation (Std) of noise component $J(t)$ for precipitation.

| Station | N | $\mu$ | Std |
|---|---|---|---|
| Biržai | 5254 | 1.006 | 1.246 |
| Dotnuva | 4615 | 1.001 | 1.221 |
| Kaunas | 5003 | 1.004 | 1.248 |
| Kybartai | 4775 | 1.007 | 1.267 |
| Klaipėda | 5024 | 1.003 | 1.170 |
| Laukuva | 5089 | 0.975 | 1.208 |
| Nida | 4656 | 1.000 | 1.108 |
| Panevėžys | 4960 | 1.003 | 1.243 |
| Raseiniai | 5485 | 0.999 | 1.191 |
| Šiauliai | 5133 | 1.007 | 1.271 |
| Šilutė | 4733 | 0.996 | 1.083 |
| Telšiai | 5064 | 1.000 | 1.245 |
| Ukmergė | 4914 | 1.000 | 1.166 |
| Utena | 5045 | 1.002 | 1.185 |
| Varėna | 4790 | 1.002 | 1.163 |
| Vilnius | 4952 | 0.999 | 1.207 |

the amount of precipitation for each month by the sum

$$\sum_{i=1}^{m} P(t_i),$$

where $t_1, \ldots, t_m$ are the consecutive days over a given month, and $m$ is the number of days in that month. This results in 1000 simulated precipitation values for each of the 12 months in the year, which we compare with the corresponding 28 (or 27) observed monthly precipitation values. Various descriptive statistics for simulated values are given in Table 3.13 together with the corresponding statistics for the observed values. Minimum and maximum values vary to some extent, however, the average observed and simulated monthly precipitation aggregates are close to each other, clearly indicating that the model describes the observed precipitation well. The 95% CIs for observed aggregated monthly precipitation are much wider than those for the corresponding simulated values. This is because the observed CIs are based on 28 data points at best, while the simulated ones are the result of 1000 outcomes. The mean observed and simulated values together with the corresponding 95% CI are plotted in Fig. 3.26.

The extremes (minimum, maximum) are well captured by the model as compared to the data. The average monthly simulated precipitation is

Table 3.13   Descriptive statistics for observed and simulated values of precipitation.

| Month | | Mean | 95% CI | Min | Max |
|---|---|---|---|---|---|
| January | Observed | 47.7 | (39.7; 55.8) | 17.3 | 99.8 |
| | Simulated | 47.9 | (47.1; 48.7) | 17.7 | 96.8 |
| Februrary | Observed | 36.9 | (31.2; 42.7) | 10.3 | 77.0 |
| | Simulated | 41.3 | (40.6; 42.1) | 8.7 | 81.0 |
| March | Observed | 41.8 | (33.5; 50.0) | 10.5 | 92.7 |
| | Simulated | 42.8 | (42.0; 43.7) | 4.6 | 115.2 |
| April | Observed | 47.9 | (39.4; 56.) | 10.5 | 96.1 |
| | Simulated | 42.7 | (41.7; 43.7) | 8.7 | 106.8 |
| May | Observed | 48.4 | (40.9; 55.9) | 14.5 | 91.8 |
| | Simulated | 54.2 | (53.0; 55.4) | 11.2 | 119.8 |
| June | Observed | 68.5 | (58.4; 78.6) | 18.5 | 127.1 |
| | Simulated | 69.5 | (68.1; 71.0) | 11.0 | 164.1 |
| July | Observed | 87.7 | (67.3; 108.2) | 11.2 | 208.7 |
| | Simulated | 83.4 | (81.7; 85.1) | 23.3 | 199.7 |
| August | Observed | 72.5 | (59.3; 85.8) | 16.0 | 148.8 |
| | Simulated | 80.5 | (78.8; 82.2) | 19.6 | 173.1 |
| September | Observed | 72.1 | (58.4; 85.8) | 11.2 | 148.1 |
| | Simulated | 64.7 | (63.4; 66.1) | 14.7 | 159.5 |
| October | Observed | 53.7 | (39.1; 68.3) | 4.3 | 149.7 |
| | Simulated | 56.9 | (55.7; 58.0) | 10.4 | 121.9 |
| November | Observed | 46.8 | (40.2; 53.4) | 10.2 | 84.9 |
| | Simulated | 49.7 | (48.8; 50.6) | 14.9 | 122.3 |
| December | Observed | 55.2 | (48.8; 61.7) | 19.2 | 85.1 |
| | Simulated | 48.8 | (48.0; 49.6) | 17.6 | 101.8 |

within the 95% CI based on observed precipitation for all month, indicating that the model describes data well.

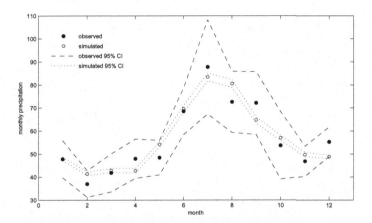

Fig. 3.26   Observed and simulated monthly precipitation values together with the corresponding 95% CI for Vilnius.

# PART 2
# Weather derivatives

# Chapter 4

# Continuous-time models for temperature and wind speed

Here we introduce the basics of the continuous-time ARMA models for temperature and wind speed. These stochastic processes are then identified with the discrete-time ARMA dynamics fitted to Lithuanian data in Chapter 3.

## 4.1 CARMA models

We introduce here the continuous-time version of the ARMA models considered for weather modelling in Chapter 3. This class of models is called *continuous-time autoregressive moving-average*, or CARMA for short, and was first introduced by [Doob (1944)]. We base our presentation of these processes on [Brockwell (2001)], who introduced the CARMA processes to financial applications.

Let $B$ be a Brownian motion defined on a *complete filtered probability space* $(\Omega, \mathcal{F}, \{\mathcal{F}_t\}_{t \geq 0}, P)$. We introduce the stochastic process $\mathbf{X}(t)$ with values in $\mathbb{R}^p$ for $p \geq 1$ as the solution of the stochastic differential equation

$$d\mathbf{X}(t) = A\mathbf{X}(t)\,dt + \mathbf{e}_p \sigma(t)\,dB(t), \tag{4.1}$$

where $\mathbf{e}_k \in \mathbb{R}^p$, $k = 1, \ldots, p$, is the $k$th standard Euclidean basis vector of $\mathbb{R}^p$. The $(p \times p)$-matrix $A$ is given by

$$A = \begin{bmatrix} 0 & 1 & 0 & \cdots & 0 \\ 0 & 0 & 1 & \cdots & 0 \\ \cdot & \cdot & \cdot & \cdot & \cdot \\ \cdot & \cdot & \cdot & \cdot & \cdot \\ 0 & 0 & 0 & 0 & 1 \\ -\alpha_p & -\alpha_{p-1} & -\alpha_{p-2} & \cdots & -\alpha_1 \end{bmatrix}. \tag{4.2}$$

The constants $\alpha_k, k = 1, \ldots, p$ are assumed to be non-negative with $\alpha_p > 0$. Recall from the empirical analysis of temperature in Sect. 3.2 that the volatility had a seasonal pattern. In order to model it, we consider a bounded and continuous function $\sigma : \mathbb{R}_+ \mapsto \mathbb{R}$ in the dynamics of $\mathbf{X}$, where $\sigma(t)$ is strictly bounded away from zero, i.e., there exists a constant $\overline{\sigma} > 0$ such that $\sigma(t) \geq \overline{\sigma}$ for all $t \geq 0$. In the sequel, we use the notation $I_n$ for the $n \times n$ identity matrix. Sometimes we simply write $I$ if the dimension is clear from the context. The following Lemma proves that $A$ is invertible.

**Lemma 4.1.** *The matrix $A$ in* (4.2) *is invertible.*

**Proof.** By the Leibnitz formula, the determinant of the matrix $A$ can be written as
$$\det(A) = -\alpha_p \det(I) = -\alpha_p.$$
Since $\alpha_p > 0$ by assumption, $\det(A) < 0$. Hence, $\det(A) \neq 0$ and the result follows. $\square$

One can compute the inverse of $A$ explicitly, as shown in the following Lemma.

**Lemma 4.2.** *The matrix $A$ has the inverse*
$$A^{-1} = \begin{bmatrix} -\dfrac{\alpha_{p-1}}{\alpha_p} & -\dfrac{\alpha_{p-2}}{\alpha_p} & \cdots & -\dfrac{\alpha_1}{\alpha_p} & -\dfrac{1}{\alpha_p} \\ 1 & 0 & \cdots & 0 & 0 \\ 0 & 1 & \cdots & 0 & 0 \\ \cdot & \cdot & \cdot & \cdot & \cdot \\ \cdot & \cdot & \cdot & \cdot & \cdot \\ 0 & 0 & \cdots & 1 & 0 \end{bmatrix}. \tag{4.3}$$

**Proof.** The proof goes by a direct verification. To simplify notation, introduce the vectors $\mathbf{a} = (\alpha_{p-1}, \alpha_{p-2}, \ldots, \alpha_1)' \in \mathbb{R}^{p-1}$ and $\mathbf{c} = (-\alpha_{p-1}/\alpha_p, -\alpha_{p-2}/\alpha_p, \ldots, -\alpha_1/\alpha_p)' \in \mathbb{R}^{p-1}$ and note that $\alpha_p \mathbf{c} + \mathbf{a} = \mathbf{0}$ with $\mathbf{0} \in \mathbb{R}^{p-1}$ being the vector of zeros. We write
$$A = \begin{bmatrix} \mathbf{0} & I_{p-1} \\ -\alpha_p & -\mathbf{a}' \end{bmatrix} \quad \text{and} \quad A^{-1} = \begin{bmatrix} \mathbf{c}' & -1/\alpha_p \\ I_{p-1} & \mathbf{0} \end{bmatrix}.$$
Then, by applying the laws of block matrix multiplication, we find
$$AA^{-1} = \begin{bmatrix} \mathbf{0}\mathbf{c}' + I_{p-1} & -\mathbf{0}/\alpha_p + I_{p-1}\mathbf{0} \\ -\alpha_p \mathbf{c}' - \mathbf{a}'I_{p-1} & 1 - \mathbf{a}'\mathbf{0} \end{bmatrix} = I_p$$
and
$$A^{-1}A = \begin{bmatrix} \mathbf{c}'\mathbf{0} + 1 & \mathbf{c}'I_{p-1} + \mathbf{a}'/\alpha_p \\ I_{p-1}\mathbf{0} - \alpha_p\mathbf{0} & I_{p-1} - \mathbf{0}\mathbf{a}' \end{bmatrix} = I_p.$$
The result follows. $\square$

In order to state the explicit solution of (4.1), the multidimensional Itô Formula is useful. For the convenience of the reader, we state this classical result from stochastic analysis in the next Theorem. A proof can be found, for example, in Thm. 4.2.1 of [Øksendal (1998)].

**Theorem 4.1.** *Assume* $\mathbf{B}(t) \in \mathbb{R}^m$ *is a Brownian motion, and let* $\mathbf{Z}(t)$ *be a stochastic process in* $\mathbb{R}^n$ *with dynamics*

$$d\mathbf{Z}(t) = \mathbf{U}(t)\, dt + V(t)\, d\mathbf{B}(t)\,.$$

*Here,* $\mathbf{U}(t) \in \mathbb{R}^n, V(t) \in \mathbb{R}^{n \times m}$ *are* $\mathcal{F}_t$*-adapted and such that*

$$P\left( \int_0^t |\mathbf{U}_i(s)|\, ds < \infty \text{ for all } t \geq 0 \right) = 1$$

*and*

$$P\left( \int_0^t |V_{ij}(s)|^2\, ds < \infty \text{ for all } t \geq 0 \right) = 1\,,$$

*for all* $i = 1, \ldots, n, j = 1, \ldots, m$. *Then for a* $C^{1,2}$ *function* $f(t,x) = (f_1(t,x), \ldots, f_r(t,x))$ *from* $[0, \infty) \times \mathbb{R}^n$ *into* $\mathbb{R}^r$ *we find*

$$df_k(t, \mathbf{Z}(t)) = \frac{\partial f_k}{\partial t}(t, \mathbf{Z}(t))\, dt + \sum_{i=1}^n \frac{\partial f_k}{\partial z_i}(t, \mathbf{Z}(t))\, d\mathbf{Z}_i(t)$$

$$+ \frac{1}{2} \sum_{i,j=1}^n \frac{\partial^2 f_k}{\partial z_i \partial z_j}(t, \mathbf{Z}(t)) d\mathbf{Z}_i(t) d\mathbf{Z}_j(t)$$

*for* $k = 1, \ldots, r$. *Here,* $d\mathbf{B}_i d\mathbf{B}_j = \delta_{ij}\, dt$ *and* $dt d\mathbf{B}_i = 0$.

We recall that a Brownian motion $\mathbf{B}(t)$ in $\mathbb{R}^m$ is simply a vector of $m$ independent Brownian motions. Furthermore, a function $f(t,x)$ is said to be $C^{1,2}$ if it is once continuously differentiable in its first argument and twice continuously differentiable in its second. The following Lemma states the explicit dynamics of $\mathbf{X}(t)$.

**Lemma 4.3.** *The solution of* (4.1) *starting at time* $t \geq 0$ *is given by the stochastic process* $\mathbf{X}(s), s \geq t$,

$$\mathbf{X}(s) = \exp(A(s-t))\mathbf{X}(t) + \int_t^s \exp(A(s-u))\mathbf{e}_p \sigma(u)\, dB(u)\,.$$

***Proof.*** The result follows from an application of the multidimensional Itô Formula in Thm. 4.1 using the function $f(s, \mathbf{x}) = \exp(As)\mathbf{x}$ with $\mathbf{x} = \mathbf{X}(s)$, as defined in (4.1). $\qquad\square$

For $0 \leq q < p$, define the vector $\mathbf{b} \in \mathbb{R}^p$ with coefficients $b_j$, $j = 0, 1, \ldots, p - 1$, satisfying $b_q = 1$ and $b_j = 0$ for $q < j < p$. We define the CARMA$(p, q)$ process as

$$Y(t) = \mathbf{b}'\mathbf{X}(t). \tag{4.4}$$

We can then express $Y(s)$ given $\mathbf{X}(t)$ for $s \geq t \geq 0$ as

$$Y(t) = \mathbf{b}' \exp(A(s - t))\mathbf{X}(t) + \int_t^s \mathbf{b}' \exp(A(s - u))\mathbf{e}_p \sigma(u)\, dB(u), \tag{4.5}$$

by applying Lemma 4.3. It is evident that $Y(s)$ conditioned on $\mathbf{X}(t)$ is a Gaussian process, with mean

$$\mathbb{E}\left[Y(s) \,|\, \mathbf{X}(t)\right] = \mathbf{b}' \exp(A(s - t))\mathbf{X}(t), \tag{4.6}$$

and variance

$$\mathrm{Var}\left(Y(s) \,|\mathbf{X}(t)\right) = \int_t^s \left(\mathbf{b}' \exp(A(s - u))\mathbf{e}_p\right)^2 \sigma^2(u)\, du. \tag{4.7}$$

A special case of the CARMA$(p, q)$ process is $q = 0$, when $\mathbf{b} = \mathbf{e}_1$, the so-called CAR$(p)$ process.

Introduce the characteristic polynomials of $Y(t)$ as

$$p(z) = z^p + \alpha_1 z^{p-1} + \cdots + \alpha_p \tag{4.8}$$

$$q(z) = b_0 + b_1 z + \cdots + b_q z^q. \tag{4.9}$$

Note that the eigenvalues of the matrix $A$ are the roots of the equation $(-1)^p p(\lambda) = 0$, or equivalently, the roots of $p(\lambda) = 0$ for the polynomial $p$ defined in (4.8). One may represent the CARMA process informally via a stochastic differential equation using the characteristic polynomials. Letting $D$ be the differential operator, we can write

$$p(D)Y(t) = q(D)DB(t).$$

This corresponds to the discrete-time difference equation for an ARMA process. In continuous-time, this representation only makes sense at an informal level, since Brownian motion has paths which are not differentiable.

We next discuss stationarity of the process $Y(t)$. Let us first restrict our attention to the classical case of constant volatility, that is, when $\sigma(t) := \sigma > 0$. If the eigenvalues of the matrix $A$ have negative real part, and the characteristic polynomials $p(z)$ and $q(z)$ in (4.8)-(4.9) do not have any common roots, the process $Y(t)$ becomes stationary (see [Brockwell (2001)]). We have from (4.6) that the mean of $Y(t)$ in stationarity (i.e., when $s \to \infty$) will be zero. The variance in stationarity becomes

$$\lim_{s \to \infty} \mathrm{Var}(Y(s) \,|\, \mathbf{X}(t)) = \sigma^2 \int_0^\infty \left(\mathbf{b}' \exp(Au)\mathbf{e}_p\right)^2 du,$$

by using (4.7). Since $Y(t)$ is Gaussian for each time $t$, and it has a limiting mean and variance, the stationary distribution of $Y$ is then also Gaussian. In the rest of this book we shall deal with matrices $A$ which fulfill such stationarity condition.

As an example, let us consider the case $p = 2$. Then the matrix $A$ becomes

$$A = \begin{bmatrix} 0 & 1 \\ -\alpha_2 & -\alpha_1 \end{bmatrix}.$$

The eigenvalues of this matrix are

$$\lambda_{1,2} = -\frac{1}{2}\alpha_1 \pm \frac{1}{2}\alpha_1 \sqrt{1 - \frac{4\alpha_2}{\alpha_1^2}},$$

which are complex when $4\alpha_2 > \alpha_1^2$, and real otherwise. Since both $\alpha_1$ and $\alpha_2$ are positive, we find that $0 \leq 1 - 4\alpha_1/\alpha_2^2 < 1$ when $4\alpha_2 < \alpha_1^2$. Thus the eigenvalues are negative in the real case, and have negative real parts in the complex case. In conclusion, for the case $p = 2$ the eigenvalues of the matrix $A$ have negative real part, and the dynamics will be stationary.

We next analyze the situation with time-dependent volatility $\sigma(t)$. In relation to weather models, we observe that $\sigma(t)$ is a periodic function, with a typical periodicity of one year. Furthermore, we can assume it to be continuous, strictly positive and bounded when modelling weather variables. We next prove that under these assumptions, the CARMA process $Y(t)$ is still stationary.

**Proposition 4.1.** *Assume $\sigma(t)$ is a function with a periodicity of $\tau > 0$, that is, $\sigma(t + \tau) = \sigma(t)$ for all $t \geq 0$. Then, if $A$ satisfies the assumptions for stationarity, $Y(t)$ has a Gaussian stationary distribution with mean zero and variance given as the limit*

$$\lim_{t \to \infty} Var(Y(t)) = \lim_{t \to \infty} \int_0^t (\mathbf{b} \exp(Au)\mathbf{e}_p')^2 \sigma^2(t - u)\, du, \qquad (4.10)$$

*which exists.*

**Proof.** It is simple to see from (4.6) that $Y(t)$ has a mean zero when $t \to \infty$. We prove that the variance of $Y(t)$ has a limit.

From the assumptions on $A$ it holds that

$$\int_0^\infty g(u)\, du < \infty$$

with $g(u) = (\mathbf{b}' \exp(Au)\mathbf{e}_p)^2$. This integral is the stationary variance of $Y(t)$ if the volatility $\sigma(t) = 1$. Introduce now the sequence

$$x_k := \int_{(k-1)\tau}^{k\tau} g(u)\sigma^2(k\tau - u)\,du,$$

for $k = 1, 2, \ldots$, and observe that the sequence

$$y_k := \sum_{i=1}^{k} x_i,$$

is such that

$$y_k = \int_0^{k\tau} g(u)\sigma^2(k\tau - u)\,du,$$

due to the periodicity of $\sigma(t)$. Each $x_k$ is positive by the assumption on $\sigma(t)$, and thus $y_k$ is a monotonely increasing sequence. By boundedness of $\sigma(t)$, we find

$$y_k \leq \sup_{t \geq 0} \sigma^2(t) \int_0^{\infty} g(u)\,du < \infty.$$

Hence, $y_k$ is a bounded monotone sequence, which therefore has a limit. This proves the Proposition. $\qquad\square$

Unless otherwise stated, we will exclusively work with stationary CARMA processes in this book. This means that we assume that $A$ satisfies the conditions of stationarity and that $\sigma(t)$ is periodic.

## 4.2 Simulation of CARMA processes

In many situations it is necessary to simulate the weather dynamics, which essentially means simulating a CARMA$(p, q)$ process. For example, we may wish to price complex weather derivatives where no analytical pricing formula exists, or we have a portfolio of weather derivatives for which we want to measure the risk. Both these cases require Monte Carlo simulations of the future weather conditions, and in this Section we derive two simple recursive schemes for this purpose. The first scheme is based on the stochastic differential equation in (4.1) describing the dynamics of the CARMA process, and the second one is based on the analytical solution of $\mathbf{X}(t)$ in Lemma 4.3. We use the idea of an Euler discretization for both schemes.

Suppose time runs on a uniform grid of step size $\Delta$, that is, $t = 0, \Delta, 2\Delta, \ldots$. From [Kloeden and Platen (1992)], we find an Euler approximation of the stochastic differential equation (4.1) given by the $p$-dimensional time series $\mathbf{x}(t) = (x_1(t), \ldots, x_p(t))'$ solving the difference equation

$$\mathbf{x}(t + \Delta) - \mathbf{x}(t) = A\mathbf{x}(t)\Delta + \mathbf{e}_p \sigma(t)\Delta B(t),$$

or, spelled out coordinate-wise,

$$x_p(t + \Delta) - x_p(t) = -\sum_{i=1}^{p} \alpha_{p-i+1} x_i(t)\Delta + \sigma(t)\Delta B(t),$$

$$x_i(t + j\Delta) - x_i(t) = x_{i+1}(t)\Delta, i = 1, \ldots, p - 1,$$

where $\Delta B(t) = B(t + \Delta) - B(t)$. We note that in distribution, $\Delta B(t) = \sqrt{\Delta}\epsilon(t)$, with $\{\epsilon(t)\}_{t=0,\Delta,2\Delta,\ldots}$ being standard normal *i.i.d.* This gives a simple recursive scheme for simulating $\mathbf{X}(t)$ at discrete times $t$.

One may increase the accuracy by looking at the analytical expression for $\mathbf{X}(t)$ as given in Lemma 4.3. We find, for times $t = 0, \Delta, 2\Delta, \ldots$,

$$\mathbf{X}(t + \Delta) = \exp(A\Delta)\mathbf{X}(t) + \int_t^{t+\Delta} \exp(A(t + \Delta - u))\mathbf{e}_p \sigma(u) \, dB(u).$$

The stochastic integral becomes a $p$-dimensional Gaussian random variable at each time step $t$. Defining $\mathbf{z}(t)$ as

$$\mathbf{z}(t) = \int_t^{t+\Delta} \exp(A(t + \Delta - u))\mathbf{e}_p \sigma(u) \, dB(u), \qquad (4.11)$$

we have that $\mathbf{z}(t)$ is normally distributed with mean zero and variance-covariance matrix given by

$$\mathrm{cov}(\mathbf{z}(t)) = \int_t^{t+\Delta} \sigma^2(u) \left( \exp(A(t + \Delta - u))\mathbf{e}_p \mathbf{e}_p' \exp(A'(t + \Delta - u)) \right) \, du$$

$$= \int_0^{\Delta} \sigma^2(t + \Delta - v) \left( \exp(Av)\mathbf{e}_p \mathbf{e}_p' \exp(A'v) \right) \, dv.$$

Furthermore, by the independent increment property of Brownian motion, $\{\mathbf{z}(t)\}, t = 0, \Delta, 2\Delta, \ldots$, are independent random variables. This yields the simulation scheme for $\mathbf{X}(t)$ given by the time series $\mathbf{x}(t)$

$$\mathbf{x}(t + \Delta) = \exp(A\Delta)\mathbf{x}(t) + \mathbf{z}(t). \qquad (4.12)$$

It requires the simulation of the independent Gaussian random vectors $\mathbf{z}(t)$. Note that $\mathbf{x}(t) = \mathbf{X}(t)$ at the discrete time points $t = 0, \Delta, 2\Delta, \ldots$, and we therefore have a scheme which exactly simulates the path of the CARMA process at given discrete times.

In the case of a constant volatility $\sigma$, we can make the properties of the covariance matrix of $\mathbf{z}(t)$ more explicit. In this case $\operatorname{cov}(\mathbf{z}(t))$ becomes independent of $t$, i.e.,

$$\operatorname{cov}(\mathbf{z}(t)) = \sigma^2 \int_0^\Delta \exp(Av)\mathbf{e}_p\mathbf{e}_p' \exp(A'v)\, dv\,.$$

To this end, let $\Sigma_\Delta := \operatorname{cov}(\mathbf{z}(t))$. We then find the following result.

**Lemma 4.4.** *The covariance matrix $\Sigma_\Delta$ of $\mathbf{z}(t)$ in the case of a constant volatility $\sigma$ is the solution of the matrix equation*

$$\sigma e^{A\Delta}\mathbf{e}_p\mathbf{e}_p'e^{A'\Delta} - \sigma\mathbf{e}_p\mathbf{e}_p' = A\Sigma_\Delta + \Sigma_\Delta A'\,.$$

**Proof.** A direct differentiation shows that

$$\frac{d}{dv}\left(e^{Av}\mathbf{e}_p\mathbf{e}_p'e^{A'v}\right) = Ae^{Av}\mathbf{e}_p\mathbf{e}_p'e^{A'v} - e^{Av}\mathbf{e}_p\mathbf{e}_p'e^{A'v}A'\,.$$

Integrating both sides from 0 to $\Delta$ with respect to $v$ yields the result.  $\square$

By taking the square root of the matrix $\Sigma_\Delta$, we can define the Euler scheme

$$\mathbf{x}(t+\Delta) = e^{A\Delta}\mathbf{x}(t) + \Sigma_\Delta^{1/2}\varepsilon(t)\,,$$

where $\varepsilon(t)$ is a vector of $p$ independent standard normally distributed random walks.

It may be slightly inefficient to simulate $\mathbf{z}(t)$ in general, as the variance-covariance matrix is time-dependent. One simple solution is to approximate the stochastic integral by

$$\int_t^{t+\Delta} \exp(A(t+\Delta-u)\mathbf{e}_p\sigma(u)\, dB(u) \approx \exp(A\Delta)\mathbf{e}_p\sigma(t)\Delta B(t)\,.$$

By using the definition of $\epsilon(t)$ above, we find an approximative Euler scheme for $\mathbf{X}(t)$ given recursively as the time series

$$\mathbf{x}(t+\Delta) = \exp(A\Delta)\left(\mathbf{x}(t) + \mathbf{e}_p\sigma(t)\sqrt{\Delta}\epsilon(t)\right)\,. \tag{4.13}$$

In this case $\mathbf{x}(t)$ only provides an approximation of $\mathbf{X}(t)$. The error will be a function of $\Delta$ as analyzed in [Kloeden and Platen (1992)].

## 4.3   Linking CARMA to ARMA

In this Section we deal with the problem of linking the CARMA model to the discrete-time series dynamics of temperature and wind. In particular, we focus on how to relate a CARMA process to a discrete-time ARMA. In effect, the results of this Section enable us to identify the parameters of a CARMA process from the estimates of an ARMA, which we found in Sect. 3 for temperature and wind speed in Lithuania.

As a first, rather straightforward, approach, we study the use of the Euler approximating scheme on the CARMA dynamics. We focus on the CAR models only, that is, on the process $Y(t) = X_1(t)$ for $X_1(t)$ being the first coordinate of the vector $\mathbf{X}(t)$ in (4.1). Applying the Euler discretization of the stochastic differential equation in (4.1) with a uniform time step given as $\Delta$ discussed in Sect. 4.2, we find a time series approximation $\mathbf{x}(t)$ for $t = 0, \Delta, 2\Delta....$, recursively defined as

$$x_p(t + \Delta) - x_p(t) = -\sum_{i=1}^{p} \alpha_{p-i+1} x_i(t)\Delta + \sigma(t)\sqrt{\Delta}\epsilon(t),$$

$$x_i(t + j\Delta) - x_i(t + (j-1)\Delta) = x_{i+1}(t + (j-1)\Delta)\Delta,$$

where $i = 1, \ldots, p-1$ and $j \geq 1$ is a natural number. Furthermore, $\epsilon(t)$ are standard normal *i.i.d.* random variables. We apply Lemma 10.2 in [Benth, Šaltytė Benth and Koekebakker (2008)] and find that

$$x_{i+1}(t) = \frac{1}{\Delta^i}\sum_{k=0}^{i}(-1)^k b_k^i x_1(t + (i-k)\Delta), \tag{4.14}$$

where the coefficients $b_k^i$ are defined as $b_0^i = b_i^i = 1$ for $i = 0, \ldots, p$ and through the recursion

$$b_k^i = b_{k-1}^{i-1} + b_k^{i-1}, k = 1, \ldots, p-1, i \geq 2. \tag{4.15}$$

Further, it is easy to show that

$$x_p(t + \Delta) - x_p(t) = \frac{1}{\Delta^{p-1}}\sum_{k=0}^{p}(-1)^p b_k^p x_1(t + (p-k)\Delta). \tag{4.16}$$

These relations yield a link between a continuous-time AR($p$) model and its discrete-time counterpart.

By using the relations above in the Euler scheme we get,

$$\frac{1}{\Delta^{p-1}} \sum_{k=0}^{p} (-1)^k b_k^p x_1(t + (p-k)\Delta)$$

$$= -\sum_{i=1}^{p} \alpha_{p-i+1} \frac{1}{\Delta^{i-2}} \sum_{k=0}^{i-1} (-1)^k b_k^{i-1} x_1(t + (i-1-k)\Delta)$$

$$+ \sigma(t)\sqrt{\Delta}\epsilon(t). \tag{4.17}$$

Observe that we have an equation which is a linear combination of $x_1(t + p\Delta), \ldots, x_1(t)$ and an error term $\sigma(t)\sqrt{\Delta}\epsilon(t)$. Thus, we find an AR($p$) time series model for $x_1(t)$. From data analysis we can estimate the coefficients of this time series process, and then use (4.17) to find the corresponding parameters $\alpha_1, \ldots, \alpha_p$ in the CAR($p$) dynamics.

Consider the example of an CAR(2) process, that is, the case of $p = 2$. By using the relation (4.17) we find that

$$\frac{1}{\Delta} \left( b_0^2 x_1(t + 2\Delta) - b_1^2 x_1(t + \Delta) + b_2^2 x_1(t) \right)$$

$$= -\alpha_2 \Delta b_0^0 x_1(t) - \alpha_1 \left( b_0^1 x_1(t + \Delta) - b_1^1 x_1(t) \right) + \sigma(t)\sqrt{\Delta}\epsilon(t).$$

Using the definition of the coefficients $b_k^i$, we have

$$x_1(t + 2\Delta) = (2 - \alpha_1\Delta)x_1(t + \Delta) + (\alpha_1\Delta - \alpha_2\Delta^2 - 1)x_1(t) + \Delta^{3/2}\sigma(t)\epsilon(t).$$

Frequently time is measured on a daily scale which yields $\Delta = 1$. This gives AR(2) model

$$x_1(t + 2) = (2 - \alpha_1)x_1(t + 1) + (\alpha_1 - \alpha_2 - 1)x_1(t) + \sigma(t)\epsilon(t).$$

To verify our formula in the trivial case of an Ornstein-Uhlenbeck process, consider $p = 1$ to find

$$b_0^1 x_1(t + \Delta) - b_1^1 x_1(t) = -\alpha_1 \frac{1}{\Delta^{-1}} b_0^0 x_1(t) + \sigma(t)\sqrt{\Delta}\epsilon(t)$$

or

$$x_1(t + \Delta) - x_1(t) = -\alpha_1 x_1(t)\Delta + \sigma(t)\sqrt{\Delta}\epsilon(t),$$

which is (not surprisingly) exactly what we would expect from the Euler scheme we started with.

One may apply the exact Euler scheme for the CARMA dynamics to link to the discrete case. This is done in [Brockwell, Davis and Yang (2007)], and goes as follows.

Let $y(n)$ be a time series of discrete observations of the CARMA process $Y(t)$ at times $n\Delta$, $n = 0, 1, \ldots$, i.e., $y(n) = Y(n\Delta)$. Then one may represent $y(n)$ as (see [Brockwell, Davis and Yang (2007)], Prop. 3)

$$\prod_{i=1}^{p}(1 - e^{\lambda_i \Delta}\mathcal{S})y(n) = \epsilon(n),$$  (4.18)

where $\mathcal{S}$ is the back-shift operator, i.e., $\mathcal{S}y(n) = y(n-1)$. Furthermore, the residuals $\epsilon(n)$ are given as

$$\epsilon(n) = \sum_{i=1}^{p} \kappa_i \prod_{j \neq i}(1 - e^{\lambda_j \Delta}\mathcal{S}) \int_{(n-1)\Delta}^{n\Delta} e^{\lambda_i(n\Delta-u)}\, dB(u),$$  (4.19)

with $\kappa_i = q(\lambda_i)/p'(\lambda_i)$, where $\lambda_1, \ldots, \lambda_p$ are the eigenvalues of $A$.

The left-hand side of (4.18) can be written in terms of an AR model for $y(n)$, namely

$$\prod_{i=1}^{p}(1-e^{\lambda_i \Delta}\mathcal{S})y(n) = y(n)+\beta_1 y(n-1)+\beta_2 y(n-2)+\cdots+\beta_p y(n-p).$$  (4.20)

The $\exp(-\lambda_i \Delta)$ is a root of the AR polynomial

$$\beta(z) = 1 - \beta_1 z - \cdots - \beta_p z^p.$$  (4.21)

We use this to estimate the parameters of the matrix $A$ in a CARMA model by first applying standard techniques for estimating $\beta_1, \ldots, \beta_p$, from observations of $y(n)$. Next, we find the distinct roots, $z_1, \ldots, z_p$, of the polynomial $\beta(z)$ in (4.21) and recover the eigenvalues of $A$ as

$$\lambda_i = -\log(z_i)/\Delta, \, i = 1, \ldots, p.$$

Since the eigenvalues are roots of the polynomial $p(z)$ in (4.8), we find $p$ linear equations for the $p$ unknowns $\alpha_1, \ldots, \alpha_p$ from $p(\lambda_i) = 0$. By solving this system we obtain the parameters of the $A$ matrix.

Let us consider a simple example of $p = 2$ and $q = 0$, that is where $Y(t)$ is a CAR(2) process. We find the polynomials $p$ and $q$ to be $p(z) = z^2 + \alpha_1 z + \alpha_2$ and $q(z) = 1$. The left-hand side of (4.18) then becomes

$$y(n) = \left(e^{\lambda_1 \Delta} + e^{\lambda_2 \Delta}\right) y(n - 1) - e^{(\lambda_1+\lambda_2)\Delta}y(n - 2) + \epsilon(n).$$

Hence, we find that $\beta_1 = \exp(\lambda_1 \Delta)+\exp(\lambda_2 \Delta)$ and $\beta_2 = -\exp((\lambda_1+\lambda_2)\Delta)$. Given the coefficients $\beta_1$ and $\beta_2$, we compute the eigenvalues $\lambda_1$ and $\lambda_2$ from the roots of the polynomial $\beta(z)$ in (4.21). The coefficients $\alpha_1$ and $\alpha_2$ are recovered by solving the linear system

$$\lambda_1 \alpha_1 + \alpha_2 = -\lambda_1^2,$$
$$\lambda_2 \alpha_1 + \alpha_2 = -\lambda_2^2.$$

We find the solution $\alpha_1 = -(\lambda_1 + \lambda_2)$ and $\alpha_2 = \lambda_1\lambda_2$. The coefficents $\kappa_i$ in (4.19) are given as $\kappa_i = 1/(2\lambda_i + \alpha_1)$, for $i = 1, 2$, where the eigenvalues $\lambda_1$ and $\lambda_2$ are roots of $p(z)$. From the definition of the residual terms we have

$$\epsilon(n) = \int_{(n-1)\Delta}^{n\Delta} \left( \kappa_1 e^{\lambda_1(n\Delta - u)} + \kappa_2 e^{\lambda_2(n\Delta - u)} \right) dB(u)$$

$$- \int_{(n-2)\Delta}^{(n-1)\Delta} \left( \kappa_1 e^{\lambda_2\Delta} e^{\lambda_1((n-1)\Delta - u)} + \kappa_2 e^{\lambda_1\Delta} e^{\lambda_2((n-1)\Delta - u)} \right) dB(u).$$

For further analysis of the relation between CARMA processes and their discrete-time analogue, we refer to [Brockwell, Davis and Yang (2007)].

## 4.4    Recovering the states I: the Kalman filter

In this section we discuss how to apply the Kalman filter to recover the states of $\mathbf{X}(t)$ from observations of a CARMA process $Y(t)$.

The problem is as follows. We are given observations of the CARMA process $Y(t)$, and want to recover the states $\mathbf{X}(t)$. To put this into a continuous-time Kalman filter, introduce the observation process

$$Z(t) = \int_0^t Y(t)\, dt. \tag{4.22}$$

Assume that we observe with noise so that the observation process has the dynamics

$$dZ(t) = Y(t)\, dt + \eta\, dW(t), \tag{4.23}$$

for a Brownian motion $W$ independent of $B$. The constant $\eta > 0$ describes the noise level in $Y$. Letting $\mathcal{Z}_t$ be the filtration generated by $Z(s), 0 \leq s \leq t$, we are searching for the best prediction of $\mathbf{X}(t)$ given $\mathcal{Z}_t$, that is,

$$\widehat{\mathbf{X}}(t) := \mathbb{E}\left[\mathbf{X}(t) \mid \mathcal{Z}_t\right]. \tag{4.24}$$

The solution to this conditional expectation is given by the Kalman filter in continuous-time.

Using Thm. 6.3.1 in [Øksendal (1998)], the solution to the filter problem is given by the stochastic differential equation

$$d\widehat{\mathbf{X}}(t) = \left(A - \eta^{-2}S(t)\mathbf{b}'\mathbf{b}\right) \widehat{\mathbf{X}}(t)\, dt + \eta^{-2}S(t)\mathbf{b}\, dZ(t), \tag{4.25}$$

with initial condition $\widehat{\mathbf{X}}(0) = \mathbb{E}[\mathbf{X}(0)]$. Here, $S(t)$ is a symmetric $(p \times p)$-matrix solving the Riccati matrix-valued differential equation

$$\frac{dS(t)}{dt} = AS(t) + S(t)A' - \eta^{-2}S(t)\mathbf{bb}'S(t) + \sigma(t)\mathbf{e}_p\mathbf{e}_p', \tag{4.26}$$

with initial value $S(0) = \text{Var}(\mathbf{X}(0))$.

It may be advantageous to re-parametrize the matrix-function $S(t)$. Introducing $S_\eta(t) := \eta^{-2}S(t)$, the filter equation becomes

$$d\widehat{\mathbf{X}}(t) = \left(A - S_\eta(t)\mathbf{b}'\mathbf{b}\right)\widehat{\mathbf{X}}(t)\,dt + S_\eta(t)\mathbf{b}\,dZ(t)\,, \qquad (4.27)$$

where $S_\eta$ solving the matrix-equation

$$\frac{dS_\eta(t)}{dt} = AS_\eta(t) + S_\eta(t)A' - S_\eta(t)\mathbf{bb}'S_\eta(t) + \eta^{-2}\sigma(t)\mathbf{e}_p\mathbf{e}_p'\,, \qquad (4.28)$$

with the initial condition $S_\eta(0) = \eta^{-2}\text{Var}(\mathbf{X}(0))$. The difference between the equation (4.26) for $S(t)$ and (4.28) for $S_\eta(t)$ is that we have moved the scaling $\eta^2$ from the second-order term to the forcing term, which has a stabilizing effect on the filter.

It is natural to use the stationary mean and variance of $\mathbf{X}(t)$ for the initial conditions of $\widehat{\mathbf{X}}(t)$ and $S_\eta(t)$, that is, we assume that

$$\widehat{\mathbf{X}}(0) = 0 \qquad (4.29)$$

and

$$S_\eta(0) = \lim_{t\to\infty}\int_0^t \exp(Au)\mathbf{e}_p\mathbf{e}_p'\exp(Au)\sigma^2(t - u)\,du\,. \qquad (4.30)$$

Strictly speaking, we do not know if this limit of $(p \times p)$-matrices exists. However, from the proof of Prop. 4.1, the limit of the variance of $\mathbf{x}'\mathbf{X}(t)$ exists for arbitrary vectors $\mathbf{x} \in \mathbb{R}^p$, whenever $\sigma(t)$ is periodic. This implies that the limit of the diagonal elements of the matrix on the right-hand side of (4.30) has a limit. Suppose now that one of the off-diagonal elements of the right-hand side of (4.30) does not have a limit. By appropriately choosing different $\mathbf{x}$ vectors, we obtain a contradiction as the limit of $\mathbf{x}'\mathbf{X}(t)$ exists. Hence, we conclude that the limit in (4.30) exists as well.

Let us present a recursive procedure to recover the states of $\mathbf{X}(t)$ based on discrete observations of $Y(t)$. The solution of $\widehat{\mathbf{X}}(t)$ in (4.27) is given by

$$\widehat{\mathbf{X}}(s) = \exp\left(\int_t^s C(u)\,du\right)\widehat{\mathbf{X}}(t) + \int_t^s \exp\left(\int_u^s C(v)\,dv\right)\mathbf{c}(u)\,dZ(u)\,,$$

for $s \geq t$, where the $(p \times p)$-matrix $C$ is defined as

$$C(t) = A - S_\eta(t)\mathbf{bb}'\,, \qquad (4.31)$$

and the $p$-dimensional vector $\mathbf{c}(t)$ as

$$\mathbf{c}(t) = S_\eta(t)\mathbf{b}\,. \qquad (4.32)$$

Considering discrete uniform times $t = 0, \Delta, 2\Delta, \ldots$, we find that

$$\widehat{\mathbf{X}}(t + \Delta) = \exp\left(\int_t^{t+\Delta} C(u)\,du\right)\widehat{\mathbf{X}}(t)$$

$$+ \int_t^{t+\Delta} \exp\left(\int_u^{t+\Delta} C(v)\,dv\right)\mathbf{c}(u)\,dZ(u).$$

Next, approximating the integrals, we find a sequence $\widehat{\mathbf{x}}(t)$ defined recursively as

$$\widehat{\mathbf{x}}(t + \Delta) = \exp(C(t)\Delta)\left(\widehat{\mathbf{x}}(t) + \mathbf{c}(t)\Delta Z(t)\right), \tag{4.33}$$

with $\Delta Z(t) = Z(t + \Delta) - Z(t)$. Using the definition of $Z$ as the integral of $Y$, we apply the approximation

$$\Delta Z(t) \approx Y(t)\Delta. \tag{4.34}$$

This provides us with an updating scheme for recovering the states of the CARMA process from observation of $Y$. Note that $\widehat{\mathbf{x}}(t)$ is an approximation of $\widehat{\mathbf{X}}(t)$.

We consider an example with discretization $\Delta = 1$ where we observe an exact simulation of a CAR(3) process $Y(t)$. An exact simulation is obtained by using the scheme (4.12), where we sample on a daily basis. The matrix $A$ is chosen to be

$$A = \begin{bmatrix} 0 & 1 & 0 \\ 0 & 0 & 1 \\ -0.187 & -1.311 & -2.034 \end{bmatrix}. \tag{4.35}$$

Such a matrix will have parameters relevant for temperatures observed in Lithuania, and, as we will see later is related to the temperature dynamics in Vilnius. For simplicity, we choose the volatility $\sigma$ to be a constant equal to one. In Figs. 4.1-4.3 we plot the filtered states (dotted lines) together with the exact states of $\mathbf{X}(t)$. The filter seems to smoothen the data considerably. This is even more evident for the filter over a month only (see Figs. 4.4-4.6).

The overall impression is that the Kalman filter captures the level and the trend of the factors, but fails on the fluctuations around them as it is much smoother. A major drawback of the Kalman filter is that it is not able to recover $Y(t) = \mathbf{X}_1(t)$ exactly. In the computations, we have applied an Euler discretization to solve numerically the involved Riccati equation, which of course introduces a numerical error.

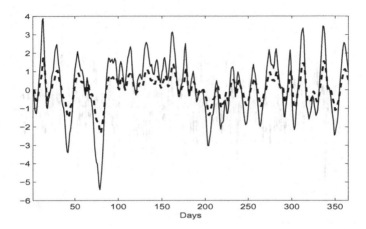

Fig. 4.1   The Kalman filtered time series $x_1(n)$ (dotted line) together with the daily sampled $\mathbf{X}_1(t)$ from an exact simulation.

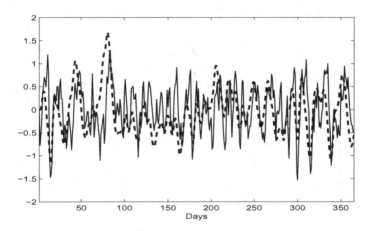

Fig. 4.2   The Kalman filtered time series $x_2(n)$ (dotted line) together with the daily sampled $\mathbf{X}_2(t)$ from an exact simulation.

## 4.5   Recovering the states II: an approxmative $L^1$-filter

In [Benth et al. (2010)] an $L^1$ filter for a CARMA$(p, q)$ process was suggested. This filter is based on an approximation of the residual process in the CARMA$(p, q)$ dynamics, and is very simple in its derivation and prac-

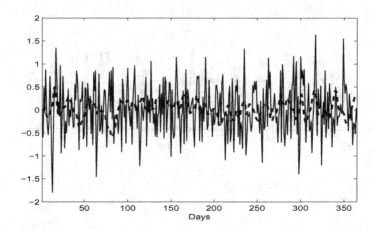

Fig. 4.3   The Kalman filtered time series $x_3(n)$ (dotted line) together with the daily sampled $X_3(t)$ from an exact simulation.

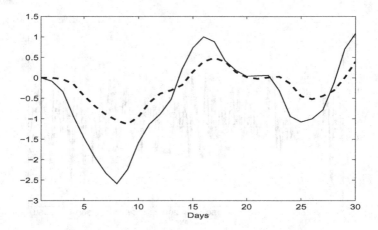

Fig. 4.4   The Kalman filtered time series $x_1(n)$ (dotted line) together with the daily sampled $X_1(t)$ from an exact simulation over a month.

tical application. To this end, let the observations from a CARMA$(p, q)$ process $Y(t)$ be denoted by $y(n)$, $n = 1, 2, \ldots$. By the exact Euler discretization discussed in Sect. 4.2, we have

$$y(n) = \mathbf{b}'\mathbf{x}(n), \tag{4.36}$$

$$\mathbf{x}(n) = \mathrm{e}^{A\Delta}\mathbf{x}(n-1) + \mathbf{z}(n), \tag{4.37}$$

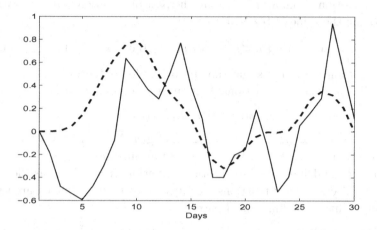

Fig. 4.5  The Kalman filtered time series $\mathbf{x}_2(n)$ (dotted line) together with the daily sampled $\mathbf{X}_2(t)$ from an exact simulation over a month.

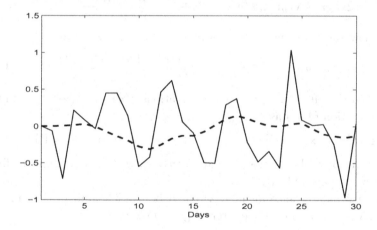

Fig. 4.6  The Kalman filtered time series $\mathbf{x}_3(n)$ (dotted line) together with the daily sampled $\mathbf{X}_3(t)$ from an exact simulation over a month.

with

$$\mathbf{z}(n) = \int_{(n-1)\Delta}^{n\Delta} e^{A(n\Delta-u)} \mathbf{e}_p \sigma(u) \, dB(u) \,. \tag{4.38}$$

If we condition on knowing $y(n)$ and the past observation $\mathbf{x}(n-1)$, we can determine the value of $\mathbf{b'z}(n)$ by

$$\mathbf{b'z}(n)\big|y(n),\mathbf{x}(n-1) = y(n) - \mathbf{b'}e^{A\Delta}\mathbf{x}(n-1). \qquad (4.39)$$

Let us for a moment investigate an approximation of the stochastic integral $\mathbf{z}(n)$. If $\Delta$ is small, it is reasonable to make the approximation

$$\mathbf{z}(n) \approx \mathbf{g}(n,\Delta)\Delta B(n),$$

where $\Delta B(n) = B(n\Delta) - B((n-1)\Delta)$ and $\mathbf{g}(n,\Delta)$ is an approximation of the function $\exp(A(n\Delta - u))\mathbf{e}_p\sigma(u)$ over the time interval $[(n-1)\Delta, n\Delta]$. Following the definition of stochastic integration, it is defined as the limit of sums of elementary functions (see [Øksendal (1998)]), which argues in favour of this approximation. Hence,

$$\mathbb{E}[\mathbf{z}(n)\,|\,y(n),\mathbf{x}(n-1)] \approx \mathbb{E}[\Delta B(n)/\Delta\,|\,y(n),\mathbf{x}(n-1)]\mathbf{g}(n,\Delta). \qquad (4.40)$$

After multiplying the expression in (4.40) with $\mathbf{b'}$ and applying equation (4.39), we get

$$\mathbb{E}[\Delta B(n)/\Delta\,|\,y(n),\mathbf{x}(n-1)] \approx \frac{y(n) - \mathbf{b'}e^{A\Delta}\mathbf{x}(n-1)}{\mathbf{b'g}(n,\Delta)}. \qquad (4.41)$$

If we now subsitute the expression (4.41) into (4.40) we obtain

$$\mathbb{E}[\mathbf{z}(n)\,|\,y(n),\mathbf{x}(n-1)] \approx \mathbf{g}(n,\Delta)\frac{y(n) - \mathbf{b'}e^{A\Delta}\mathbf{x}(n-1)}{\mathbf{b'g}(n,\Delta)}. \qquad (4.42)$$

We reach the filter for $\mathbf{x}(n)$ given observations $y(n)$ and $\mathbf{x}(n-1)$ from the state equation (4.37) as

$$\mathbb{E}[\mathbf{x}(n)\,|y(n),\mathbf{x}(n-1)] = e^{A\Delta}\mathbf{x}(n-1) + \mathbb{E}[\mathbf{z}(n)\,|\,y(n),\mathbf{x}(n-1)]. \qquad (4.43)$$

Hence, given an observation $y(n)$ and $\mathbf{x}(n-1)$, we find $\mathbf{x}(n)$ from the filter equations (4.42) and (4.43).

A reasonable choice for $\mathbf{g}(n,\Delta)$ could be the left-limit of the function $\exp(A\Delta - u)\mathbf{e}_p\sigma(u)$ over the interval $[(n-1)\Delta, n\Delta]$, that is,

$$\mathbf{g}(n,\Delta) = e^{A\Delta}\mathbf{e}_p\sigma((n-1)\Delta). \qquad (4.44)$$

Alternatively, as chosen in [Benth et al. (2010)], we can take $\mathbf{g}(n,\Delta)$ to be the average value over the interval. If $\sigma(u) = \sigma$ is a constant, we find that

$$g(n,\Delta) = \frac{1}{\Delta}\int_{(n-1)\Delta}^{n\Delta} e^{A(n\Delta-u)}\mathbf{e}_p\sigma\,du = -\frac{\sigma}{\Delta}A^{-1}(I - e^{A\Delta})\mathbf{e}_p. \qquad (4.45)$$

For a time-dependent volatility $\sigma$ we may not be able to obtain an analytical expression for the average. However, numerical integration can be carried out for this purpose. Also, if $\sigma(u)$ is not varying too much over the interval, one can first approximate its value by some suitable constant (the average, say), and next take the average using this constant as the proxy for the volatility.

It is noteworthy that if we simulate a path of the CARMA$(p, q)$ dynamics using the semi-exact scheme in (4.13), and apply the choice of $\mathbf{g}(n, \Delta)$ in (4.44), the proposed $L^1$-filter will be able to recover the states exactly. The reason for this is the fact that all the approximations in the derivations above are found with equality, since we simulate from a model of $\mathbf{X}(t)$ where the noise is, in fact, given by $\mathbf{g}(n, \Delta)\Delta B(n)$ and not by $\mathbf{z}(n)$ itself. It is an advantageous property of the filter.

However, in practice we must assume that the data are sampled from a CARMA$(p, q)$ model, and not coming from any approximate numerical sampling scheme. In order to have a test on the quality of the filter, we can simulate a path of the CARMA$(p, q)$ process $Y(t)$ using the exact scheme in (4.12). Given samples from this path, we can try to recover $\mathbf{x}(n)$ using the filter based on a certain choice of $\mathbf{g}(n, \Delta)$. In the specification of a numerical example, we choose a constant volatility $\sigma = 1$ and a time discretization length of $\Delta = 1$. The matrix $A$ is assumed to be as in (4.35). We assume furthermore $\mathbf{b}' = (1, 0, 0)$, such that $Y(t)$ is a CAR(3) process with constant volatility. The vector-valued function $\mathbf{g}(n, 1)$ was chosen as the average defined in (4.45).

In Fig. 4.7, we show the simulated path of $Y(t)$ sampled daily over a year. The filter recovers this factor exactly. The filtered factors $\mathbf{x}_2(n)$ and $\mathbf{x}_3(n)$ are shown in Figs. 4.8-4.9 (dotted lines) together with the exact values (solid lines). The filter behaves reasonably well and captures the overall variations. The filter performs better for the second factor as compared with the third. The error is of course attributed to the sampling approximations in the filter procedure.

To see better the performance of the filter, we have zoomed the results over a month of 30 days, as seen in Figs. 4.10-4.11. It is clear that the factor $\mathbf{X}_2(t)$ is very well recovered. For the last factor, $\mathbf{X}_3(t)$, the filter is not as good. However, we still see that it recovers the basic pattern and level of variations. We conclude that the proposed $L^1$-filter is perfoming very well. This combined with a simple implementation makes it very attractive in applications. Compared to the Kalman filter, the $L^1$-filter seems to work far better.

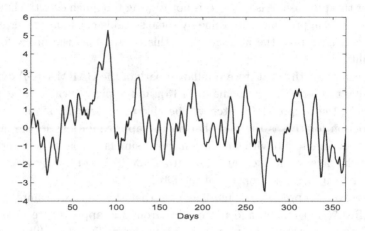

Fig. 4.7   A simulation of the process $Y(t)$ sampled daily.

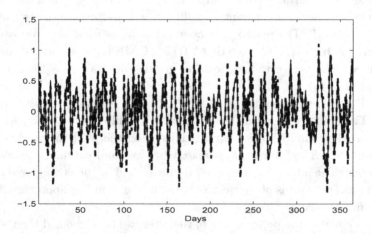

Fig. 4.8   The filtered time series $x_2(n)$ (dotted line) together with the daily sampled values of $X_2(t)$ from an exact simulation.

## 4.6   CARMA models for temperature and wind speed

We present CARMA models for the dynamics of temperature and wind speed measured in one single location. Our modelling is motivated by and based on the discrete time analysis in Sect. 3.2.

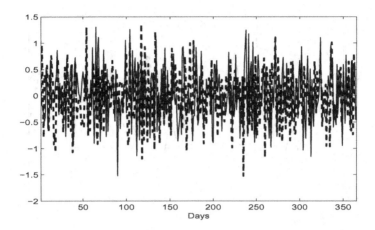

Fig. 4.9   The filtered time series $x_3(n)$ (dotted line) together with the daily sampled values of $X_3(t)$ from an exact simulation.

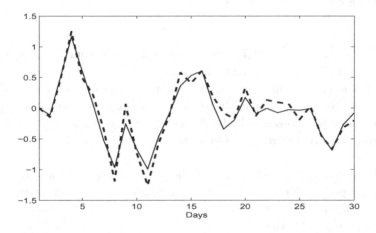

Fig. 4.10   The filtered time series $x_2(n)$ (dotted line) together with the daily sampled values of $X_2(t)$ from an exact simulation over a month.

### 4.6.1   *A model for temperature*

As we recall from the empirical analysis of temperature data, an AR process with seasonal component was fitting data well. We formulate this model in a continuous-time framework, using the CARMA class introduced above.

Assume that the temperature $T(t)$ at time $t \geq 0$, follows a CARMA$(p, q)$

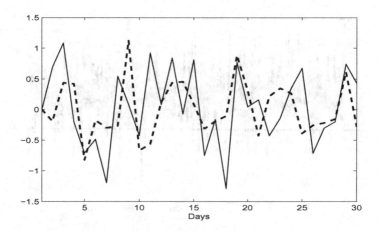

Fig. 4.11   The filtered time series $\mathbf{x}_3(n)$ (dotted line) together with the daily sampled values of $\mathbf{X}_3(t)$ from an exact simulation over a month.

dynamics for $q < p$, that is,

$$T(t) = \Lambda(t) + Y(t), \tag{4.46}$$

with $Y(t) = \mathbf{b}'\mathbf{X}(t)$ and $\Lambda(t)$ being a continuous and bounded seasonality function. Although we argued for an AR(3) structure in our empirical analysis of temperatures in Sect. 3.2, we work with the more general CARMA processes here. We note in passing that this class of processes have been proposed and applied successfully by [Garcia, Klüppelberg and Müller (2010)] for modelling the dynamics of electricity spot prices (see also [Benth et al. (2010)]).

As long as the matrix $A$ satisfies the conditions for stationarity of the CARMA process $Y(t)$ and the volatility is periodic, we observe that the *deseasonalized* temperatures $T(t) - \Lambda(t)$ are stationary. From (4.6) and stationarity it holds that

$$\lim_{t \to \infty} \mathbb{E}[T(t)] - \Lambda(t) = 0.$$

Or, in other words, the long term level of temperature is simply the seasonal function $\Lambda(t)$. Thus, the CARMA$(p, q)$ model implies that temperature mean-reverts to the seasonal level $\Lambda(t)$, the average temperature at time $t$.

Recall now the empirical findings for temperature data from Lithuania presented in Sect. 3.2. In Table 3.1, the estimated parameters of the trend and seasonal component function in (3.10) can be found. Following the empirical analysis, we choose

$$\Lambda(t) = a_0 + a_1 t + a_2 \cos(2\pi t/365) + a_3 \sin(2\pi t/365) \, .$$

The AR(3) process was found to fit the deseasonalized temperature best, with estimated parameters in Table 3.1. A natural choice for $Y(t)$ is therefore a CAR(3) process, that is, $Y(t) = e_3' \mathbf{X}(t)$ with $p = 3$. Using relation (4.17) for $p = 3$ and $\Delta = 1$ yields

$$x_1(t+3) = (3 - \alpha_1) x_1(t+2) + (2\alpha_1 - \alpha_2 - 3) x_1(t+1)$$
$$+ (\alpha_2 + 1 - \alpha_1 - \alpha_3) x_1(t) + \sigma(t)\epsilon(t) \, .$$

Assessing Table 3.1 and taking Vilnius as the case, we get the linear system of equations

$$3 - \alpha_1 = 0.966 \, ,$$
$$2\alpha_1 - \alpha_2 - 3 = -0.243 \, ,$$
$$\alpha_2 + 1 - \alpha_1 - \alpha_3 = 0.090 \, ,$$

with the following solution

$$\alpha_1 = 2.034 \, , \quad \alpha_2 = 1.311 \quad \alpha_3 = 0.187 \, .$$

We observe that all the $\alpha_i$, $i = 1, 2, 3$, are positive, just as assumed. The eigenvalues of the corresponding matrix $A$ are then

$$\lambda_1 = -0.197 \, , \quad \lambda_{2,3} = -0.9185 \pm 0.3247i \, ,$$

all with a negative real part. This means that the CAR(3) process is of a stationary type, which is in line with the stationarity of the AR(3) discrete-time series estimated for Vilnius. The volatility function $\sigma(t)$ is chosen according to the specification in (3.11). In Table 3.2 the estimates for the parameters are presented for $L_\sigma = 4$ in (3.11). This gives a full specification of the temperature model in a continuous time framework. In Fig. 4.12 we plot simulated temperature over five years in Vilnius based on the fitted model. In the simulation we applied the semi-exact scheme developed in (4.13). The time step was chosen to be $\Delta = 1$, and we started the recursion at January 1 with the states $\mathbf{X}$ equal to zero, implying that the initial temperature were set equal to its mean.

## 4.6.2 A model for wind speed

We recall from the empirical analysis of wind speed in Sect. 3.2 that an ARMA model performed well on logarithmically transformed data. Thus,

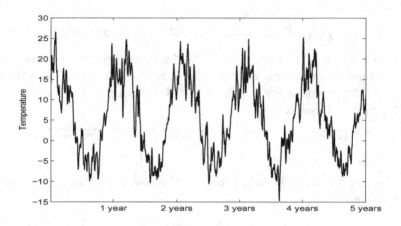

Fig. 4.12   A simulation of the daily temperature in Vilnius.

it is natural to consider an exponential CARMA$(p, q)$ dynamics for the wind speed in continuous-time. However, we would like to be slightly more general in the modelling of wind speed, as the analysis of New York data in [Šaltytė Benth and Benth (2010)] suggested that a power transform is more suitable than the logarithmic. By using a Box-Cox power transform, [Šaltytė Benth and Benth (2010)] symmetrized the wind speed data, and fitted an AR(4) process with seasonal mean and volatility. Since the Box-Cox transform also includes the logarithmic function, we consider this general transformation of wind speed data here when defining a continuous-time dynamics.

Referring to the Box-Cox transform, define the *wind speed* $W(t)$ at time $t \geq 0$ to be

$$W(t) = \begin{cases} (\lambda(\Lambda(t) + Y(t)) + 1)^{1/\lambda}, & \lambda \neq 0, \\ \exp(\Lambda(t) + Y(t)), & \lambda = 0, \end{cases} \qquad (4.47)$$

for a constant $\lambda$, where $\Lambda(t)$ is a continuous and bounded deterministic seasonal function. Again, even though the empirical analysis suggests an AR(4) model, which would naturally be interpreted as $Y(t)$ being a CARMA(4,0), we keep the general CARMA$(p, q)$ framework here as we did for temperature. Hence, assume that $Y(t) = \mathbf{b}'\mathbf{X}(t)$.

Observe that the Box-Cox transformed wind speed for $\lambda \neq 0$ is

$$\frac{1}{\lambda} \left( W^\lambda(t) - 1 \right) = \Lambda(t) + Y(t),$$

and for $\lambda = 0$

$$\ln W(t) = \Lambda(t) + Y(t).$$

Hence, we have a seasonal CARMA$(p, q)$ dynamics of transformed wind speed. It is interesting to note that by choosing $p = 1$ and $q = 0$, we recover the so-called Schwartz dynamics used for modelling the spot price of oil (see [Schwartz (1997)]). This geometric Ornstein-Uhlenbeck process have later been extended and applied in electricity markets (see [Benth, Šaltytė Benth and Koekebakker (2008)] for an extensive analysis).

A problem with the Box-Cox transform is that for $\lambda \neq 0$ we may theoretically get negative wind speed. Of course, this is avoided in the exponential ($\lambda = 0$) case, and whenever $1/\lambda$ turns out to be an even number. The empiricial analysis in Sect. 3.2 used the logarithmically transformed wind speed, so this is not an issue for the data collected in Lithuania. However, going back to the study of New York wind speed in [Šaltytė Benth and Benth (2010)], it was shown that $1/\lambda = 5$ performed best. Thus, since the CARMA process $Y(t)$ can attain any negative value with positive probability, one may potentially have negative wind speed in the stochastic dynamics. On the other hand, with the estimated model parameters for New York (see [Šaltytė Benth and Benth (2010)]), the probability of such negative values is very low, and negligible in practice.

The empirical analysis in Sect. 3.2 suggests the seasonal function of Lithuanian wind speed data on logarithmic scale to be of the form

$$\Lambda(t) = a_0 + a_2 \cos(2\pi t/365) + a_3 \sin(2\pi t/365).$$

The notable difference from the temperature model is that the trend were found to be insignificant for wind speed. The estimated parameters of the seasonal function for the various locations are presented in Table 3.6. In the same table the estimated regression coefficients in the AR model for the deseasonalized logarithmically transformed wind speeds are given. Seven of the locations are best modelled by an AR(4) time series dynamics. In five locations, an AR(3) seems to be the preferred choice, the remaining four locations should be modelled using an AR(6) or AR(8). From the temperature case, we know how to identify the analogous CAR(3) model, so we focus here on the situation $p = 4$ (leaving the higher-order cases to the interested reader). Using relation (4.17) for $p = 4$ and $\Delta = 1$ yields

$$
\begin{aligned}
x_1(t+4) = {} & (4 - \alpha_1)x_1(t+3) + (3\alpha_1 - \alpha_2 - 6)x_1(t+2) \\
& + (4 + 2\alpha_2 - \alpha_3 - 3\alpha_1)x_1(t+1) \\
& + (\alpha_3 - \alpha_4 - \alpha_2 + \alpha_1 - 1)x_1(t) + \sigma(t)\epsilon(t).
\end{aligned}
$$

Taking Biržai as the case in Table 3.6, the estimated AR coefficients yield the system of equations for $\alpha_1, \ldots, \alpha_4$

$$4 - \alpha_1 = 0.46\,,$$
$$3\alpha_1 - \alpha_2 - 6 = -0.05\,,$$
$$4 + 2\alpha_2 - \alpha_3 - 3\alpha_1 = 0.05\,,$$
$$\alpha_3 - \alpha_4 - \alpha_2 + \alpha_1 - 1 = 0.02\,.$$

Solving this system gives us the coefficients in the $A$ matrix of the CAR(4) dynamics of Biržai to be $\alpha_1 = 3.54, \alpha_2 = 4.67, \alpha_3 = 2.67$ and $\alpha_4 = 0.52$. All the four $\alpha$ values are positive, as required in the definition of the CAR($p$) process. The eigenvalues of the matrix $A$ then become $\lambda_1 = -1.2163, \lambda_{2,3} = -0.9641 \pm 0.3894i$ and $\lambda_4 = -0.3954$. The real parts are all negative, giving a stationary CAR(4) dynamics. This is in line with the AR(4) model estimated for Biržai, where the discrete-time series is stationary. The volatility function $\sigma(t)$ can be chosen according to the specification in (3.11), or as a constant. The analysis of wind speed data argued in favour of a constant volatility, with the estimate in Table 3.7 (first column). If one wants to use a seasonal volatility function, $L_\sigma = 1$ in the wind speed case, the parameters of the function (3.11) can also be found in Table 3.7. This provides us with a full specification of the wind speed model in continuous-time.

The dynamics of wind speed is very important for wind mill power generators. The energy content of wind passing through a wind mill is defined as

$$E(t) = \frac{1}{2}\rho a W^3(t)\,. \tag{4.48}$$

Here, $a$ is the area spanned by the rotor blades, and the constant $\rho$ is the density of the air. The density can differ according to temperature, humidity etc. From Betz' law (see [Betz (1966)]), only a part of this power can actually be extracted for energy production due to energy losses. The maximum part is theoretically the fraction $16/27$, or $59.3\%$. Modern wind mills can produce energy close to this upper limit. The power produced from a wind mill is therefore proportional to the third power of wind speed, where the proportionality constant $c$ is close to $16 \times \rho \times a/54$. Using our model for wind speed, we see that the power, $P(t)$, produced at time $t$, is given as

$$P(t) = c \left(\lambda(\Lambda(t) + Y(t)) + 1\right)^{3/\lambda}\,, \tag{4.49}$$

whenever $\lambda \neq 0$, while for $\lambda = 0$ it is

$$P(t) = x \exp\left(\Lambda(t) + Y(t)\right)\,. \tag{4.50}$$

Our CARMA model can therefore be used to predict the power production from a wind mill.

## 4.7 Speed of reversion to the mean: the half-life

It is often of interest to understand how fast the process is reverting back to its mean. Of course, considering a deterministic dynamics like

$$\frac{dx(t)}{dt} = -\alpha x(t) \,,$$

we know that the speed of $x(t)$ at each time $t$ is $-\alpha x(t)$ since the velocity of the dynamics is given by its derivative. But when adding a stochastic term to obtain an Ornstein-Uhlenbeck process, or more generally, a CARMA process, is there any sensible way to measure how fast the process is returning to its mean? [Clewlow and Strickland (2000)] introduce the so-called half-life of the process, being the expected time it takes for the process to return half the way back to its mean from some state. We make this concept rigorous for CARMA processes and discuss it in relation to some concrete situations.

Recall that the stationary mean of the CARMA process $Y(t)$ is zero, so we want to measure how fast the process on average is returning back to this level. Define the *half-life* of the CARMA$(p,q)$ process $Y(t)$ as the first time $\tau > t$ such that

$$\mathbb{E}[Y(\tau) \,|\, \mathcal{F}_t] = \frac{1}{2}Y(t)\,. \tag{4.51}$$

We observe that $\tau$ will be dependent on current time $t$, and stochastic. By the properties of conditional expectation, $\tau$ will be a stopping time.

From Lemma 4.3 we find that the half-life $\tau$ will be the solution of the equation

$$\mathbf{b}' \exp(A(\tau - t))\mathbf{X}(t) = \frac{1}{2}\mathbf{b}'\mathbf{X}(t)\,. \tag{4.52}$$

Hence, $\tau$ will be dependent on $t$ and $\mathbf{X}(t)$, and we denote it by $\tau(t, \mathbf{X}(t))$. That is, the speed of reversion to the mean of the process is state-dependent. Note that we may have several solutions, and we define the smallest one to be the half-life.

The next proposition shows that there exists a solution of (4.52).

**Proposition 4.2.** *There exists a half-life $\tau \in (t, \infty)$ being the solution of* (4.52)*.*

**Proof.** To prove the existence of the half-life, let us introduce the auxiliary real-valued function

$$f(\tau) = \mathbf{b}' \left( \exp(A(\tau - t)) - \frac{1}{2}I \right) \mathbf{X}(t)\,.$$

We observe that the half-life $\tau(t, \mathbf{X}(t))$ is a root of $f$. Clearly we have that $f(\tau)$ is a continuous function on $[t, \infty)$, and since $A$ has eigenvalues with negative real parts (from the stationarity of the CARMA process), we obtain that

$$\lim_{\tau \to \infty} f(\tau) = -\frac{1}{2} \mathbf{b}' \mathbf{X}(t).$$

On the other hand, it follows that $f(t) = \frac{1}{2} \mathbf{b}' \mathbf{X}(t)$. Thus, $f(t)$ will have the opposite sign of $\lim_{\tau \to \infty} f(\tau)$, and it follows from continuity of $f$ that it has at least one root in the interval $(t, \infty)$. The smallest of these roots will be the half-life. $\qquad \square$

Consider the simple example of an Ornstein-Uhlenbeck process, i.e. $p = 1$ and $q = 0$. Then (4.52) becomes

$$\exp(-\alpha(\tau - t))X(t) = \frac{1}{2}X(t)$$

or

$$\tau - t = \frac{\ln(2)}{\alpha}.$$

Clearly, for an Ornstein-Uhlenbeck process the half-life is independent of the state of the process. If $\alpha$ is large, then the half-life is small, meaning that the process reverts back quickly to its mean level zero. On the other hand, a small $\alpha$ gives a larger half-life, so that the process uses more time to revert back to zero. The constant $\alpha$ is determining the speed of reversion to the mean of the process. Usually, one thinks of $\alpha$ as the speed of mean reversion of an Ornstein-Uhlenbeck process.

In our next example, we have considered a CAR(3) process with the $A$ matrix given by

$$A = \begin{bmatrix} 0 & 1 & 0 \\ 0 & 0 & 1 \\ -0.177 & -1.339 & -2.043 \end{bmatrix}.$$

We remark that this choice of an $A$ matrix coincides with the one that was estimated for temperature data in Stockholm, Sweden, in [Benth, Šaltytė Benth and Koekebakker (2008)]. Moreover, we assume a constant volatility of $\sigma = 2$. In Fig. 4.13 we plot the average half-life $\tau(t, \mathbf{X}(t)) - t$ computed from a simulation of 10,000 paths of the CAR(3) process over the times $t = 1, \ldots, 20$. The states $\mathbf{X}(t) = (X_1(t), X_2(t), X_3(t))'$ were simulated using the Euler scheme introduced in (4.13). The initial state at time zero was assumed to be $\mathbf{X}(0) = (0, 0, 0)'$, and for each path $\tau(t, \mathbf{X}(t))$ was

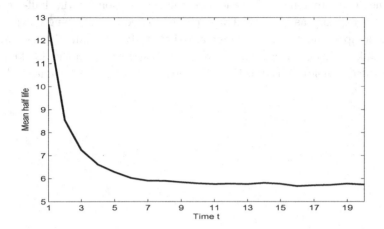

Fig. 4.13 The mean half-life $\tau(t, \mathbf{X}(t)) - t$ computed based on 10,000 sample paths of a CAR(3) process.

computed by solving (4.52). From Fig. 4.13 it is clear that the mean half-life less current time is converging to a stable level slightly below six. It decays (close to exponentially) from an initial level of around 12.5. This initial decay is due to the simulation of the CAR(3) process not starting in stationarity, but at a fixed given point, and thus the path needs some time to reach stationarity.

Let us discuss these findings for the particular CAR(3) example in relation to an Ornstein-Uhlenbeck process, a CAR(1) process. We can compute the eigenvalues of the matrix $A$ to be

$$\lambda_1 = -0.175 \,, \text{and } \lambda_{2,3} = -0.934 \pm 0.374\mathrm{i} \,.$$

As expected, they have negative real parts meaning stationarity of the CAR(3) process. Denoting the corresponding eigenvectors $\mathbf{v}_i$, $i = 1, 2, 3$, we find the half-life associated with $\mathbf{X}(t) = \mathbf{v}_i$ as the solution of

$$\mathbf{b}' \exp(A(\tau - t))\mathbf{v}_i = \frac{1}{2}\mathbf{b}'\mathbf{v}_i$$

or

$$\tau_i - t = -\frac{\ln(2)}{\lambda_i} \,.$$

Inserting the real parts of the eigenvalues, we see that the dominating term is coming from $\lambda_1$, and yields $\tau_1 - t = 3.96$. If we let $\alpha = -\lambda_1$ be the speed of mean reversion in an Ornstein-Uhlenbeck process, the half-life becomes close to four. We hence see the implication on the half-life of having a memory as in the CAR(3) process, effectively slowing down the reversion speed on average. We remark that in [Benth, Šaltytė Benth and Koekebakker (2008)] the CAR(1) process was estimated on the Stockholm temperature, yielding $\alpha_1 = 0.177$, very close to the first eigenvalue of $A$.

# Chapter 5

# Pricing of forward contracts on temperature and wind speed

This chapter discusses various "classical" approaches to pricing weather contracts written on temperature and wind speed. The approaches include the rational expectation hypothesis and the burn analysis. The latter does not provide a dynamical price, but can only give one constant price. We focus on the risk-neutral pricing approach including the rational expectation hypothesis as a special case, and price various contracts traded in the weather markets for temperature and wind speed.

## 5.1 Theory on pricing forwards

We are mainly concerned with two types of derivatives in this book, forward contracts and options on forwards. The no-arbitrage theory for derivatives pricing is based on hedging arguments.

Taking a forward contract on a tradeable commodity, the *forward price* is defined by resorting to the *buy-and-hold strategy*. For example, consider an investor selling a forward contract at time $t$ on coffee beans with delivery at time $T > t$, with the forward price $f(t,T)$. The investor must deliver the agreed amount of coffee beans at time $T$, and receive the forward price $f(t,T)$ as payment. In order to hedge this position, the seller may simply buy coffee beans today in the spot market for the current spot price $S(t)$, and store it until delivery. To finance the purchase, the seller may borrow the required amount of money in the bank, paying $r$ in compounded interest. At delivery, the seller receives $f(t,T)$, delivers the coffee beans, and pays back the loan which is the amount $S(t)\exp(r(T-t))$. If the forward price is $f(t,T) > S(t)\exp(r(T-t))$, *the seller* has made an arbitrage. If the opposite is true, that is $f(t,T) < S(t)\exp(r(T-t))$, *the buyer* of the forward contract can make an arbitrage by shorting coffee beans in the spot market. Hence, the arbitrage-free price of the forward contract on coffee beans will be

$$f(t,T) = S(t) \exp(r(T-t)) \,. \tag{5.1}$$

We note that this relationship between the forward and the spot price at time $t$ will hold as long as we can buy coffee beans and store them at no cost. Furthermore, we also assume in this pricing relation that we can enter short positions in coffee beans in the spot market. Hence, the basic assumption is that of a perfectly liquid spot market, where the underlying can be traded without frictions. The pricing rule in (5.1) is the classical spot-forward relationship, which holds without any model assumptions on the spot dynamics as long as the buy-and-hold hedging strategy can be implemented.

The general arbitrage pricing theory tells us that a market is free of arbitrage as long as there exists at least one equivalent martingale measure $Q$ (see [Duffie (1992)]). An equivalent measure $Q \sim P$ is a martingale measure as long as the discounted prices of the tradeable assets in the market are all martingales under $Q$. Assuming $S(t)$ being tradeable, we have that $\exp(-rt)S(t)$ is a $Q$-martingale, and by using the martingale property we get that

$$\begin{aligned}
f(t,T) &= S(t) \exp(r(T-t)) \\
&= S(t) \exp(-rt) \exp(rT) \\
&= \mathbb{E}_Q \left[ \exp(-rT)S(T) \,|\, \mathcal{F}_t \right] \exp(rT) \\
&= \mathbb{E}_Q \left[ S(T) \,|\, \mathcal{F}_t \right] .
\end{aligned}$$

In conclusion,

$$f(t,T) = \mathbb{E}_Q \left[ S(T) \,|\, \mathcal{F}_t \right] . \tag{5.2}$$

Here, $\mathbb{E}_Q$ is the expectation operator with respect to the probability $Q$. From this pricing relation, it follows immediately that the stochastic process $t \mapsto f(t,T)$ for $t \leq T$ is itself a $Q$-martingale. We note that if $S(t)$ is a semimartingale, one is ensured the existence of equivalent martingale measures (see e.g. [Shiryaev (1999)]).

The underlying "commodity" of a weather forward contract is an index linked to specific weather events. For example, we may have a contract which "delivers" HDD over a specified month in a year, in return for an agreed forward price. The HDD is calculated from temperatures in the given month, and can as such obviously not be traded in any liquid fashion. Hence, we cannot resort to the above no-arbitrage argument based on the buy-and-hold strategy. But, since the forward contract is itself a tradeable asset, the price dynamics must be a martingale with respect to some equivalent probability $Q$. Note that we do not demand $Q$ to be an equivalent *martingale* measure for the underlying "commodity price dynamics". This is simply so because the underlying, the HDD or CDD temperature index, say, cannot be traded.

The way to obtain a forward price which is a martingale under some equivalent measure, and at the same time links the price to the underlying weather index, is to *define* it as the conditional expectation under some $Q$ of the index. This is the standard approach, and the one usually chosen. We will apply this as the starting point for deriving the forward price dynamics for weather-linked forward contracts. The approach is analogous to the pricing of zero-coupon bonds based on the short-rate of interest in fixed income market theory.

To be slightly more specific, suppose we have a forward which "delivers" the HDD at a particular time instance $\tau$, or, in other words a forward that delivers to the buyer the money equivalent of

$$\text{HDD}(\tau) := \max(18 - T(\tau), 0),$$

where $T(\tau)$ is the temperature at time $\tau$. The forward price is then defined as

$$f(t, \tau) = \mathbb{E}_Q \left[ \text{HDD}(\tau) \,|\, \mathcal{F}_t \right], \tag{5.3}$$

for $t \leq \tau$ and some equivalent probability $Q$. We observe that the martingale property of the forward dynamics $t \mapsto f(t, \tau)$ with respect to the probability $Q$, follows from the very definition of the conditional expectation. This ensures that we have an arbitrage-free market consisting of the risk-free bank account and the forward contract. Implicitly, we assume here that the HDD is integrable with respect to $Q$[1].

In fixed income market theory, the *rational expectation hypothesis* is often used as a tool for pricing derivatives based on the interest rate, for instance, zero-coupon bonds. Translated to the weather derivatives setting, the rational expectation hypothesis would be to choose $Q = P$. Hence, one chooses a forward price which is the best prediction of the index at maturity, given the current information. This yields a zero *risk premium*,

---

[1]Physically, temperatures have a lower limit at $0°$ Kelvin, so the HDD is bounded. Also, in practical terms this is so. However, when specifying a stochastic model for the temperature, we must ensure that the HDD is integrable since, the model may typically attain all values on the real line.

defined as the difference between the forward price and the predicted index
at maturity, that is,

$$R(t, \tau) := f(t, \tau) - \mathbb{E}\left[\mathrm{HDD}(\tau) \,|\, \mathcal{F}_t\right]. \tag{5.4}$$

From a hedger's point of view, selling a forward contract to secure a certain
index value at maturity $\tau$, one could expect a willingness to accept a lower
forward price than the predicted index value. Such a situation corresponds
to *normal backwardation*, where the risk premium is negative. The forward
price is in *contango* when the risk premium is positive. Both cases (or the
combination of the two) demand a choice of $Q$ different than $P$.

### 5.1.1   *Pricing by burn analysis*

The so-called *burn analysis* is a classical approach for pricing weather
derivatives (see [Jewson and Brix (2005)]). Burn analysis simply uses the
historical distribution of the weather index/event underlying the derivative
as the basis for pricing.

For example, to find a price of an option based on the HDD index in
a given month, January say, we first collect historical HDD index values
from January in preceding years. Based on these records, we generate the
historical option payoffs, and simply price the option by averaging.

To be more precise, assume that we have temperature records for January over the years $y_1, y_2, \ldots, y_n$. Thus we can calculate a series of historical January HDD values $h_1, h_2, \ldots, h_n$. If the option is of call type, with
strike $K$, we compute the payoffs $p = \max(h - K, 0)$ to obtain the series
$p_1, p_2, \ldots, p_n$. The option price according to the burn analysis will then be
given by

$$\mathrm{Price} = \frac{1}{n} \sum_{i=1}^{n} p_i.$$

The burn analysis therefore corresponds to pricing by the historical expectation, that is, by choosing the pricing measure $Q = P$. In finance, one
often calls this type of pricing by the rational expectation hypothesis. Note,
however, the burn analysis is model-free.

Recall Sect. 3.3, where we validated our time series model for precipitation on monthly aggregated values. From a long time series of daily
precipitation values, the number of data points for the cumulative precipitation in a given month was reduced to only 28 (or 27 for some locations).
This means that applying the burn analysis to pricing a derivative on precipitation, we would rely on only 28 data points. In the situation of a call

or put option on the cumulative amount of rainfall in a given month, this could give very few non-zero payoff data, and therefore result in a highly uncertain expectation value.

Even though our focus in this Chapter is on wind speed and temperature forwards, the same critics of the burn analysis approach applies here, as most of the products offered in the market is based on monthly (or longer) aggregations of the underlying weather variable. In conclusion, relying on aggregated observations may lead to very inaccurate estimates, yielding a high degree of statistical uncertainty in the price estimate.

On the other hand, using a stochastic model for the underlying dynamics as discussed in previous Chapters, will rely on a huge amount of data, and thereby ensure a much more statistically reliable description of the weather variable in question. Hence, contrary to the burn analysis, the estimation of prices becomes reliable with only a small degree of statistical uncertainty attached to it. This is the main reason for dismissing the burn analysis as a viable method for pricing weather derivatives. We propose instead to use dynamical stochastic models for the underlying weather factor estimated to a rich set of daily data.

If we aim at finding a price *dynamics* for a weather derivative, the burn analysis is not appropriate either. Indeed, if we aim at finding the forward price of a contract delivering the HDD index for January, for example, by using the burn analysis we would simply average the numbers $h_i$ above. However, this produces a constant forward price not varying with time, so we will not obtain any forward price *dynamics* (except for a trivial constant one). The price of options on temperature forwards will not be interesting in this situation, as it will be equal to the payoff. In fact, there will not be any reason for options in the market. Hence, the burn analysis does not provide any reliable approach for pricing of temperature derivatives in a liquid weather market.

## 5.2   A structure preserving class of measure changes

We introduce a class of equivalent probabilities $Q$ to be used as the pricing measures in the pricing of weather forwards. The measures change the level of mean reversion in the CARMA dynamics $\mathbf{X}(t)$ from zero to a constant, and in addition change the AR coefficients $\alpha_1, \ldots, \alpha_p$. The latter may be interpreted as a change in the speed of mean reversion of the stationary CARMA dynamics.

Recall the definition of $\mathbf{X}(t)$ in (4.1). Define a vector $\theta = (\theta_p, \ldots, \theta_1)' \in \mathbb{R}^{p2}$, and suppose that

$$\theta_i < \alpha_i, i = 1, \ldots, p. \tag{5.5}$$

Introduce the process

$$d\widetilde{B}(t) = \sigma^{-1}(t)\left(\theta_0(t) + \theta'\mathbf{X}(t)\right) dt + dB(t), \tag{5.6}$$

for a bounded and measurable real-valued function $\theta_0(t)$. We prove that $\widetilde{B}(t)$ is a Brownian motion under some equivalent probability measure, $Q^\theta$, as long as the eigenvalues of the matrix $A$ have negative real parts (recall from Prop. 4.1 that this implies stationarity of $\mathbf{X}$ for periodic volatility).

**Proposition 5.1.** *Assume that the eigenvalues of the matrix $A$ all have negative real part. Then, the process $\widetilde{B}(t)$ in (5.6) is a Brownian motion for $t \in [0, \tau]$ for each $\tau < \infty$ with respect to the equivalent probability measure $Q^\theta$ with density process*

$$\frac{dQ^\theta}{dP}\bigg|_{\mathcal{F}_t} = \exp\left(\int_0^t \frac{\theta_0(s) + \theta'\mathbf{X}(s)}{\sigma(s)} dB(s) - \frac{1}{2}\int_0^t \frac{(\theta_0(s) + \theta'\mathbf{X}(s))^2}{\sigma^2(s)} ds\right). \tag{5.7}$$

**Proof.** The proposition follows from the Girsanov Theorem (see [Karatzas and Shreve (1991)]) as long as the density process is a martingale. The proof of this goes by an application of a *modified version* of the Novikov condition that is found in Cor. 5.14 in [Karatzas and Shreve (1991)], p. 199.

According to Cor 5.14 in [Karatzas and Shreve (1991)], we have to show that there exists a sequence of real numbers $\{t_k\}_{k=0}^\infty$, $0 = t_0 < t_1 < \ldots < t_k \uparrow \infty$ such that for all $k \geq 0$

$$\mathbb{E}\left[\exp\left(\frac{1}{2}\int_{t_k}^{t_{k+1}} \frac{(\theta_0(t) + \theta'\mathbf{X}(t))^2}{\sigma^2(t)} dt\right)\right] < \infty. \tag{5.8}$$

Recall from the definition of the CARMA process in Sect. 4.1 that $\sigma(t)$ is assumed to be lower bounded by $\overline{\sigma}$. Furthermore, since $\theta_0(t)$ is assumed to be bounded, we get

$$\int_{t_k}^{t_{k+1}} \frac{(\theta_0(t) + \theta'\mathbf{X}(t))^2}{\sigma^2(t)} dt \leq \frac{1}{\overline{\sigma}^2}\int_{t_{k-1}}^{t_k} (\theta_0(t) + \theta'\mathbf{X}(t))^2 dt$$

$$\leq \frac{\overline{\theta}_0^2}{\overline{\sigma}^2}(t_{k+1} - t_k) + \frac{2\overline{\theta}_0}{\overline{\sigma}^2}\int_{t_k}^{t_{k+1}} \theta'\mathbf{X}(t) dt$$

$$+ \frac{1}{\overline{\sigma}^2}\int_{t_k}^{t_{k+1}} (\theta'\mathbf{X}(t))^2 dt,$$

---

[2]The reader should note the unusual labelling of the coordinates in the vector $\theta$ going backwards.

with $\sup_{0 \le t < \infty} |\theta_0(t)| = \bar{\theta}_0$. Next, we find from the explicit solution of $\mathbf{X}(t)$ in Lemma 4.3 that the first integral in the above inequality becomes

$$\int_{t_k}^{t_{k+1}} \theta' \mathbf{X}(t)\, dt = \int_{t_k}^{t_{k+1}} \theta' \exp(At)\mathbf{x}\, dt + \int_{t_k}^{t_{k+1}} \int_0^t g(t-s)\sigma(s)\, dB(s)\, dt\,,$$

where

$$g(u) = \theta' \exp(Au)\mathbf{e}_p\,.$$

The second integral in the inequality can be estimated as

$$\int_{t_k}^{t_{k+1}} (\theta' \mathbf{X}(t))^2\, dt = \int_{t_k}^{t_{k+1}} (\theta' \exp(At)\mathbf{x})^2\, dt$$

$$+ 2 \int_{t_k}^{t_{k+1}} \theta' \exp(At)\mathbf{x} \int_0^t g(t-s)\sigma(s)\, dB(s)\, dt$$

$$+ \int_{t_k}^{t_{k+1}} \left( \int_0^t g(t-s)\sigma(s)\, dB(s) \right)^2\, dt$$

$$\le 2 \int_{t_k}^{t_{k+1}} (\theta' \exp(At)\mathbf{x})^2\, dt$$

$$+ 2 \int_{t_k}^{t_{k+1}} \left( \int_0^t g(t-s)\sigma(s)\, dB(s) \right)^2\, dt\,.$$

Following from the assumption of $A$ having eigenvalues with negative real part, the function $t \mapsto \theta' \exp(At)\mathbf{x}$ approaches zero as $t \to \infty$. Hence, from continuity, it follows that this function must be bounded by some constant $c > 0$. Thus, we reach the estimate

$$\int_{t_k}^{t_{k+1}} \frac{(\theta_0(t) + \theta' \mathbf{X}(t))^2}{\sigma^2(t)}\, dt \le c_1(t_{k+1} - t_k)$$

$$+ c_2 \int_{t_k}^{t_{k+1}} \int_0^t g(t-s)\sigma(s)\, dB(s)\, dt$$

$$+ c_3 \int_{t_k}^{t_{k+1}} \left( \int_0^t g(t-s)\sigma(s)\, dB(s) \right)^2\, dt\,,$$

for some positive constants $c_i$, $i = 1, 2, 3$. Hence, by applying Hölder's inequality (see [Folland (1984)]), we obtain

$$\mathbb{E}\left[\exp\left(\frac{1}{2}\int_{t_k}^{t_{k+1}}\frac{(\theta_0(t)+\theta'\mathbf{X}(t))^2}{\sigma^2(t)}\,dt\right)\right]$$

$$\le e^{c_1(t_{k+1}-t_k)}\mathbb{E}\left[\exp\left(c_2\int_{t_k}^{t_{k+1}}\int_0^t g(t-s)\sigma(s)\,dB(s)\,dt\right)\right.$$

$$\left.\times\exp\left(c_3\int_{t_k}^{t_{k+1}}\left(\int_0^t g(t-s)\sigma(s)\,dB(s)\right)^2\,dt\right)\right]$$

$$\le e^{c_1(t_{k+1}-t_k)}\mathbb{E}\left[\exp\left(c_2 p\int_{t_k}^{t_{k+1}}\int_0^t g(t-s)\sigma(s)\,dB(s)\,dt\right)\right]^{1/p}$$

$$\times\mathbb{E}\left[\exp\left(c_3 q\int_{t_k}^{t_{k+1}}\left(\int_0^t g(t-s)\sigma(s)\,dB(s)\right)^2\,dt\right)\right]^{1/q}.$$

$$(5.9)$$

The constants $p, q > 1$ are reciprocal conjugates, that is, $1/p + 1/q = 1$.

Consider the first expectation in (5.9). The stochastic Fubini Theorem (see [Protter (1990)], Thm. 45, p. 159) yields

$$\int_{t_k}^{t_{k+1}}\int_0^t g(t-s)\sigma(s)\,dB(s)\,dt$$

$$=\int_{t_k}^{t_{k+1}}\int_0^{t_k} g(t-s)\sigma(s)\,dB(s)\,dt+\int_{t_k}^{t_{k+1}}\int_{t_k}^t g(t-s)\sigma(s)\,dB(s)\,dt$$

$$=\int_0^{t_k}\int_{t_k}^{t_{k+1}} g(t-s)\,dt\sigma(s)\,dB(s)+\int_{t_k}^{t_{k+1}}\int_s^{t_{k+1}} g(t-s)\,dt\sigma(s)\,dB(s).$$

Thus, by the independent increment property of Brownian motion, the first expectation term in (5.9) can be split into two expressions for the exponential moment of a centered normal random variable. That is,

$$\mathbb{E}\left[\exp\left(c_2 p\int_{t_k}^{t_{k+1}}\int_0^t g(t-s)\sigma(s)\,dB(s)\,dt\right)\right]$$

$$=\mathbb{E}\left[\exp\left(c_2 p\int_0^{t_k}\int_{t_k}^{t_{k+1}} g(t-s)\,dt\sigma(s)\,dB(s)\right)\right]$$

$$\times\mathbb{E}\left[\exp\left(c_2 p\int_{t_k}^{t_k}\int_s^{t_{k+1}} g(t-s)\,dt\sigma(s)\,dB(s)\right)\right]$$

$$=\exp\left(\tilde{c}_1\int_0^{t_k}\left(\int_{t_k}^{t_{k+1}} g(t-s)\,dt\right)^2\,ds\right)$$

$$\times\exp\left(\tilde{c}_2\int_{t_k}^{t_{k+1}}\left(\int_s^{t_{k+1}} g(t-s)\,dt\right)^2\,ds\right),$$

for positive constants $\tilde{c}_1$ and $\tilde{c}_2$. Using the assumption on the eigenvalues of $A$, we apply the spectral representation to show that there exists a positive constant $\lambda > 0$ (the smallest real part of the eigenvalues in absolute value) such that $|g(t-s)| \leq \tilde{c}\exp(-\lambda(t-s))$ for some constant $\tilde{c} > 0$, which yields

$$\int_0^{t_k} \left( \int_{t_k}^{t_{k+1}} g(t-s)\,dt \right)^2 ds \leq K \int_0^{t_k} e^{2\lambda s} \left( e^{-\lambda t_k} - e^{-\lambda t_{k+1}} \right)^2 ds$$

$$= K \left( 1 - e^{-2\lambda t_k} \right) \left( 1 - e^{-\lambda(t_{k+1}-t_k)} \right)$$

$$\leq K,$$

where $K$ is some suitably chosen constant. For the second integral, we find that

$$\int_{t_k}^{t_{k+1}} \left( \int_s^{t_k} g(t-s)\,dt \right)^2 ds \leq K \int_{t_k}^{t_{k+1}} e^{2\lambda s} \left( e^{-\lambda s} - e^{-\lambda t_{k+1}} \right)^2 ds$$

$$= K \int_{t_k}^{t_{k+1}} \left( 1 - e^{-\lambda(t_{k+1}-s)} \right)^2 ds$$

$$\leq K(t_{k+1} - t_k).$$

In conclusion, for the first expectation term in (5.9), we are left with

$$\mathbb{E}\left[ \exp\left( c_2 p \int_{t_k}^{t_{k+1}} \int_0^t g(t-s)\sigma(s)\,dB(s)\,dt \right) \right] \leq \exp(K_1 + K_2(t_{k+1} - t_k)),$$

for two positive constants $K_1$ and $K_2$.

We move our attention to the second expectation in (5.9). It holds that

$$\mathbb{E}\left[ \exp\left( c_3 q \int_{t_{k-1}}^{t_k} \left( \int_0^t g(t-s)\sigma(s)\,dB(s) \right)^2 dt \right) \right]$$

$$= \sum_{n=0}^{\infty} \frac{(c_3 q)^n}{n!} \mathbb{E}\left[ \left( \int_{t_{k-1}}^{t_k} \left( \int_0^t g(t-s)\sigma(s)\,dB(s) \right)^2 dt \right)^n \right].$$

By Minkowski's inequality for integrals (see [Folland (1984)], p. 186),

$$\mathbb{E}\left[ \left( \int_{t_{k-1}}^{t_k} \left( \int_0^t g(t-s)\sigma(s)\,dB(s) \right)^2 dt \right)^n \right]^{1/n}$$

$$\leq \int_{t_{k-1}}^{t_k} \mathbb{E}\left[ \left( \int_0^t g(t-s)\sigma(s)\,dB(s) \right)^{2n} \right]^{1/n} dt$$

$$= \mathbb{E}[Z^{2n}]^{1/n} \int_{t_{k-1}}^{t_k} \int_0^t g^2(t-s)\sigma^2(s)\,ds\,dt$$

$$\leq \mathbb{E}[Z^{2n}]^{1/n}\hat{\sigma}^2 \int_{t_k}^{t_{k+1}} \int_0^t g^2(s)\,ds\,dt,$$

where, in distribution,

$$\int_0^t g(t-s)\sigma(s)\, dB(s) = Z \sqrt{\int_0^t g^2(t-s)\sigma^2(s)\, ds} \,,$$

for a standard normal random variable $Z$ and $\widehat{\sigma} = \sup_{0 \le t < \infty} |\sigma(s)|$. Applying again the assumption on the eigenvalues of $A$ and the spectral representation of $g$, we find,

$$\int_{t_k}^{t_{k+1}} \int_0^t g^2(s)\, ds\, dt = K \int_{t_k}^{t_{k+1}} \int_0^t e^{-2\lambda s}\, ds\, dt$$

$$= K \int_{t_k}^{t_{k+1}} (1 - e^{-2\lambda t})\, dt \le K(t_{k+1} - t_k)\,,$$

for some constant $K$ (that may have changed its value in each step). It follows that

$$\mathbb{E}\left[ \exp\left( c_q \int_{t_k}^{t_{k+1}} \left( \int_0^t g(t-s)\sigma(s)\, dB(s) \right)^2 dt \right) \right]$$

$$\le \sum_{n=0}^{\infty} \frac{1}{n!} \widetilde{K}^n (t_k - t_{k-1})^n \mathbb{E}[Z^{2n}]\,,$$

for a suitably chosen positive constant $\widetilde{K}$. Now, since

$$\mathbb{E}[Z^{2n}] = (2n-1)\mathbb{E}[Z^{2(n-1)}]\,,$$

the series above converges by the ratio test whenever

$$2\widetilde{K}(t_{k+1} - t_k) < 1.$$

Hence, by choosing the $t_k$'s such that

$$t_{k+1} - t_k < \frac{1}{2\widetilde{K}}\,,$$

we obtain a finite series. We conclude also that the second expectation in (5.9) is finite. Summing up, we showed that there exists a sequence of $\{t_k\}$ with $t_k \uparrow \infty$ such that the expectation in (5.8) is finite for every $k \ge 1$. Hence, the proof is complete. $\qquad\square$

Usually the Novikov condition (see [Karatzas and Shreve (1991)]) is applied in order to validate a Girsanov transform. However, as we can see from the estimations in the proof above, it would result in a strong condition on the time horizon $\tau$ for the validity of the Girsanov transform. In fact, this horizon strongly depends on the parameters of the problem, and we may experience situations where the measure change is only valid for very small times $\tau$. This means that we may not be able to incorporate an analysis of all weather forwards of interest as their measurement period may be beyond this horizon. The situation is mended by applying the modified Novikov condition as in the proof, allowing for a general specification of the CARMA model, time horizon $\tau$ and parameters $\theta_0$ and $\theta$.

Obviously, if we let $\theta = \mathbf{0}$, with $\mathbf{0}$ a $p$-dimensional vector of zeros, then the measure change becomes

$$d\widetilde{B}(t) = \frac{\theta_0(t)}{\sigma(t)}\, dt + dB(t),$$

and a straightforward application of the Novikov condition validates that $Q^\theta$ is a probability measure and $\widetilde{B}$ is a $Q^\theta$ Brownian motion, because $\sigma(t)$ is assumed to be bounded away from zero and $\theta_0$ is bounded. It is the changing of the $\alpha_i$ parameters of the matrix $A$ in the CARMA process that leads to the rather lengthy and cumbersome proof of Prop. 5.1 to validate the Girsanov transform. We note in passing that the vector $\theta$ may also be time-dependent, but we refrain from this generality here.

The $Q^\theta$-dynamics of $\mathbf{X}(t)$ is

$$d\mathbf{X}(t) = (\mathbf{e}_p\theta_0(t) + A_\theta \mathbf{X})\, dt + \mathbf{e}_p\sigma(t)\, d\widetilde{B}(t), \tag{5.10}$$

where

$$A_\theta = \begin{bmatrix} 0 & 1 & 0 & \cdots & 0 \\ 0 & 0 & 1 & \cdots & 0 \\ . & . & . & \cdots & . \\ . & . & . & \cdots & 1 \\ -(\alpha_p - \theta_p) & -(\alpha_{p-1} - \theta_{p-1}) & -(\alpha_{p-2} - \theta_{p-2}) & \cdots & -(\alpha_1 - \theta_1) \end{bmatrix}. \tag{5.11}$$

Recall by the condition on $\theta_i$ that all the coefficients $\alpha_i - \theta_i$, $i = 1, \ldots, p$, are positive. Here we also see the reason for the unusual labelling of the coordinates in the vector $\theta$. Thus, $\mathbf{b}'\mathbf{X}(t)$ is a CARMA model also under $Q^\theta$. In this respect we may interpret the measure change induced by (5.6) as *structure preserving*. We state now the explicit dynamics of $\mathbf{X}(t)$ under $Q^\theta$.

**Lemma 5.1.** *Under $Q^\theta$, it holds for $s \geq t \geq 0$, that*

$$\mathbf{X}(s) = \exp(A_\theta(s - t))\mathbf{X}(t) + \int_t^s \exp(A_\theta(s - u))\mathbf{e}_p\theta_0(u)\, du$$

$$+ \int_t^s \exp(A_\theta(s - u))\mathbf{e}_p\sigma(u)\, d\widetilde{B}(u).$$

**Proof.**    Apply the multidimensional Itô Formula, Thm. 4.1, for the function $f(s, \mathbf{x}) = \exp(As)\mathbf{x}$ and $\mathbf{x} = \mathbf{X}(s)$ with dynamics given as in (5.10) to complete the proof.    □

Let us consider a simple example where $p = 1$ and $\sigma(t) = \sigma$, $\theta_0(t) = \theta_0$ are constants. Then the measure yields the $Q^\theta$ Brownian motion

$$d\widetilde{B}(t) = \sigma^{-1}(\theta_0 + \theta_1 X(t))\, dt + dB(t)\,.$$

The $Q^\theta$-dynamics of $\mathbf{X}(t) = X(t)$ becomes

$$dX(t) = (\theta_0 - (\alpha_1 - \theta_1)X(t))\, dt + \sigma\, d\widetilde{B}(t)\,,$$

which is a stationary Ornstein-Uhlenbeck process under $Q^\theta$ since $\alpha_1 - \theta_1 > 0$ by assumption. In the case $p = 2$ the measure change $Q^\theta$ also preserves stationarity of the CARMA process. This can be readily seen from the example in Sect. 4.1.

## 5.3    Pricing temperature forwards

In Chapter 1 we introduced and discussed weather futures contracts written on the CAT, HDD and CDD temperature indices. The HDD index over the measurement period $[\tau_1, \tau_2]$, with $\tau_1 < \tau_2$, is defined to be the aggregation of temperatures below a threshold $c$ in the period $[\tau_1, \tau_2]$. Mathematically, it can be expressed as

$$\mathrm{HDD}(\tau_1, \tau_2) := \int_{\tau_1}^{\tau_2} \max(c - T(\tau), 0)\, d\tau\,, \tag{5.12}$$

where $T(\tau)$ is the temperature at time $\tau$ and $c$ is the threshold, usually 18°C. We recall that in the market place, the HDD index is measured discretely, aggregating the daily HDD index calculated from the daily average temperature. For mathematical convenience, we here define the HDD index by integration rather than discrete summation. The CDD index is defined accordingly as

$$\mathrm{CDD}(\tau_1, \tau_2) := \int_{\tau_1}^{\tau_2} \max(T(\tau) - c, 0)\, d\tau\,, \tag{5.13}$$

whereas the CAT index is given by

$$\mathrm{CAT}(\tau_1, \tau_2) := \int_{\tau_1}^{\tau_2} T(\tau)\, d\tau\,. \tag{5.14}$$

Consider a forward contract delivering one of the above indices. We want to study the forward price of this contract at time $t \leq \tau_1$, and define it to be

$$F_{\text{Ind}}(t, \tau_1, \tau_2) := \mathbb{E}_Q\left[\text{Ind}(\tau_1, \tau_2) \,|\, \mathcal{F}_t\right], \tag{5.15}$$

where Ind is a generic reference to the temperature index in question, with measurement period $[\tau_1, \tau_2]$. The definition of the forward price is motivated from the discussion in Sect. 5.1, where, for the moment, $Q$ is any equivalent probability used as a pricing measure. Let us study some simple relations between the different indices and forwards.

We have the following Lemma on the *CDD-HDD index parity.*

**Lemma 5.2.** *It holds that*

$$CDD(\tau_1, \tau_2) - HDD(\tau_1, \tau_2) = CAT(\tau_1, \tau_2) - c(\tau_2 - \tau_1).$$

**Proof.** Note that

$$\max(x - c, 0) - \max(c - x, 0) = x - c.$$

Integration yields the result. □

The CDD-HDD index parity implies an arbitrage relation among the three forwards on CDDs, HDDs and CATs, and is the analogue of the put-call parity in option theory. The following Corollary states the relationship among the forward prices.

**Corollary 5.1.** *It holds that,*

$$F_{CDD}(t, \tau_1, \tau_2) - F_{HDD}(t, \tau_1, \tau_2) = F_{CAT}(t, \tau_1, \tau_2) - c(\tau_2 - \tau_1).$$

**Proof.** This is proven by taking conditional expectations in the CDD-HDD index parity in Lemma 5.2. □

We now continue with the pricing of these forwards based on our model for temperature given in (4.46). As we recall, the empirical analysis on temperature data suggested a CAR($p$) dynamics, with $p = 3$. As there is no essential complication to consider a general CARMA($p, q$) dynamics, we do so in the sequel.

The pricing measure $Q$ is assumed to be $Q^\theta$ as defined in Sect. 5.2. From Lemma 5.1 we find that $T(s)$ given $\mathbf{X}(t)$ for $s \geq t$ can be represented as

$$T(s) = m_\theta(t, s, \mathbf{X}(t)) + \int_t^s \mathbf{b}' \exp(A_\theta(s - u)) \mathbf{e}_p \sigma(u) \, d\tilde{B}(u), \tag{5.16}$$

with $m_\theta(t, s, \mathbf{X}(t))$ defined by

$$m_\theta(t, s, \mathbf{X}(t)) = \Lambda(s) + \mathbf{b}' \exp(A_\theta(s-t))\mathbf{X}(t)$$
$$+ \int_t^s \mathbf{b}' \exp(A_\theta(s-u))\mathbf{e}_p\theta_0(u)\, du\,. \tag{5.17}$$

If we let

$$\Sigma_\theta^2(t, s) = \int_t^s \left(\mathbf{b}' \exp(A_\theta(s-u))\mathbf{e}_p\right)^2 \sigma^2(u)\, du\,, \tag{5.18}$$

then we can express $T(s)$ conditioned on $\mathcal{F}_t$ for $s \geq t$ with respect to $Q^\theta$ in distribution as

$$T(s) = m_\theta(t, s, \mathbf{X}(t)) + \Sigma_\theta(t, s)\, Z\,, \tag{5.19}$$

where $Z$ is a standard normally distributed random variable. The conditional representation of $T(s)$ in (5.19) will become convenient when computing temperature futures prices.

Our first result concerns CAT futures. We introduce the following notation to obtain more compact expressions. Define for $0 \leq v \leq u$ the $(p \times p)$-matrix

$$C_\theta(u, v) = A_\theta^{-1}\left(\exp(A_\theta u) - \exp(A_\theta v)\right)\,. \tag{5.20}$$

It is convenient to extend the definition of $C_\theta(u, v)$ for $v < 0$ (but $u \geq 0$) by letting

$$C_\theta(u, v) = A_\theta^{-1}\left(\exp(A_\theta u) - I\right)\,. \tag{5.21}$$

Recall from Lemma 4.1 that the matrix $A$ is invertible. Hence, using the same arguments as in the proof of Lemma 4.1, $A_\theta$ is invertible as well, since $\alpha_p - \theta_p > 0$ from the restriction on $\theta_p$. A simple modification of Lemma 4.2 yields an explicit expression for the inverse $A_\theta^{-1}$, given as the matrix in (4.3) after substituting $\alpha_i$ with $\alpha_i - \theta_i$, $i = 1, \ldots, p$.

The following holds for the CAT futures price dynamics.

**Proposition 5.2.** *For $t \leq \tau_1$ and $\tau_2 > \tau_1$, it holds that*

$$F_{CAT}(t, \tau_1, \tau_2) = \int_{\tau_1}^{\tau_2} \Lambda(s)\, ds + \mathbf{b}'C_\theta(\tau_2 - t, \tau_1 - t)\mathbf{X}(t)$$
$$+ \int_t^{\tau_2} \mathbf{b}'C_\theta(\tau_2 - s, \tau_1 - s)\mathbf{e}_p\theta_0(s)\, ds\,.$$

**Proof.**    We have from the $Q^\theta$-dynamics of $\mathbf{X}(s)$ in Lemma 5.1, that

$$F_{\mathrm{CAT}}(t, \tau_1, \tau_2) = \mathbb{E}_\theta \left[ \int_{\tau_1}^{\tau_2} T(s) \, ds \mid \mathcal{F}_t \right]$$

$$= \int_{\tau_1}^{\tau_2} \Lambda(s) \, ds + \mathbf{b}' \int_{\tau_1}^{\tau_2} \exp(A_\theta(s - t)) \, ds \mathbf{X}(t)$$

$$+ \int_{\tau_1}^{\tau_2} \int_t^s \mathbf{b}' \exp(A_\theta(s - u)) \mathbf{e}_p \theta_0(u) \, du \, ds \,,$$

where we applied the adaptedness of $\mathbf{X}(t)$ to $\mathcal{F}_t$ and the independence of the Wiener integral. Considering the last integral, we have

$$\int_{\tau_1}^{\tau_2} \int_t^s \mathbf{b}' \exp(A_\theta(s - u)) \mathbf{e}_p \theta_0(u) \, du \, ds$$

$$= \int_{\tau_1}^{\tau_2} \int_t^{\tau_1} \mathbf{b}' \exp(A_\theta(s - u)) \mathbf{e}_p \theta_0(u) \, du \, ds$$

$$+ \int_{\tau_1}^{\tau_2} \int_{\tau_1}^s \mathbf{b}' \exp(A_\theta(s - u)) \mathbf{e}_p \theta_0(u) \, du \, ds \,.$$

Applying the Fubini-Tonelli Theorem, we can commute the integrals, and reach the result of the Proposition after integrating.    □

The $Q^\theta$-dynamics of $F_{\mathrm{CAT}}(t, \tau_1, \tau_2)$ is computed in the next Proposition by using Itô's Formula.

**Proposition 5.3.** *The $Q^\theta$-dynamics of $F_{CAT}(t, \tau_1, \tau_2)$ for $t \leq \tau_1 < \tau_2$, is*

$$dF_{CAT}(t, \tau_1, \tau_2) = \mathbf{b}' C_\theta(\tau_2 - t, \tau_1 - t) \mathbf{e}_p \sigma(t) \, d\tilde{B}(t) \,.$$

**Proof.**    First, observe that $t \mapsto F_{\mathrm{CAT}}(t, \tau_1, \tau_2)$ is a $Q^\theta$-martingale and therefore it must be expressible as an Itô integral. Hence, we see from Prop. 5.2 that the only relevant term in the dynamics of $F_{\mathrm{CAT}}$ is coming from the last coordinate in the differential of $\mathbf{X}(t)$. This simplifies the usage of Itô's Formula significantly.    □

We move on with pricing CDD futures. To simplify the considerations, we introduce the following notation. Let the function $\Psi(x)$ defined on $x \in \mathbb{R}$ be given as

$$\Psi(x) = x\Phi(x) + \phi(x) \,, \tag{5.22}$$

with $\Phi(x)$ being the cumulative standard normal distribution and $\phi(x)$ its derivative, that is, the probability density function.

**Proposition 5.4.** *It holds for $t \leq \tau_1 < \tau_2$ that*

$$F_{CDD}(t, \tau_1, \tau_2) = \int_{\tau_1}^{\tau_2} \Sigma_\theta(t, s) \Psi \left( \frac{m_\theta(t, s, \mathbf{X}(t)) - c}{\Sigma_\theta(t, s)} \right) \, ds \,,$$

*with $\Psi(x)$ defined in (5.22).*

**Proof.**    Recall from (5.19) that for $T(s)$ conditional on $\mathcal{F}_t$, $s \geq t$, we have in distribution that

$$T(s) = m_\theta(t, s, \mathbf{X}(t)) + Z\Sigma_\theta(t, s),$$

for $m_\theta(t, s, \mathbf{X}(t))$ and $\Sigma_\theta(t, s)$ defined in (5.17) and (5.18) respectively, and $Z$ being a standard normally distributed random variable. Hence, by applying the Fubini-Tonelli theorem and the properties of the normal distribution, the CDD futures price becomes

$$F_{\text{CDD}}(t, \tau_1, \tau_2) = \mathbb{E}_\theta \left[ \int_{\tau_1}^{\tau_2} \max(T(s) - c, 0) \, ds \mid \mathcal{F}_t \right]$$

$$= \int_{\tau_1}^{\tau_2} \mathbb{E}_\theta \left[ \max(T(s) - c, 0) \mid \mathcal{F}_t \right] ds$$

$$= \int_{\tau_1}^{\tau_2} \mathbb{E} \left[ \max(m_\theta(t, s, \mathbf{X}(t)) + Z\Sigma_\theta(t, s) - c, 0) \right] ds$$

$$= \int_{\tau_1}^{\tau_2} \int_{y(t,s,\mathbf{X}(t))}^{\infty} (m_\theta(t, s, \mathbf{X}(t)) - c + z\Sigma_\theta(t, s)) \, \phi(z) \, dz \, ds.$$

Here, $y(t, s, \mathbf{x}) = (m_\theta(t, s, \mathbf{x}) - c)/\Sigma_\theta(t, s)$. A straightforward calculation yields the Proposition.    □

One can obtain a different expression for the CDD futures price. First, note that

$$\Psi'(x) = x\Phi'(x) + \Phi(x) + \phi'(x) = \Phi(x),$$

since $\Phi'(x) = \phi(x)$ and $\phi'(x) = -x\phi(x)$. Moreover, from L'Hopital's rule we easily see that $\Psi(x)$ tends to zero when $x \to -\infty$. Hence, by the Fundamental Theorem of Calculus, it follows that

$$\Psi(x) = \int_{-\infty}^{x} \Phi(y) \, dy = \int_{-\infty}^{x} P(Z \leq y) \, dy, \tag{5.23}$$

for a standard normally distributed random variable $Z$. Observe in passing that since $\Phi(x)$ tends to one and $\phi(x)$ tends to zero when $x \to \infty$, we have that $\Psi(x) \sim x$ for large $x$.

The CDD futures price can alternatively be expressed as follows.

**Corollary 5.2.** *The CDD futures price* $F_{CDD}(t, \tau_1, \tau_2)$ *can be represented as*

$$F_{CDD}(t, \tau_1, \tau_2)$$

$$= \int_{\mathbb{R}} P(Z \leq y) \int_{\tau_1}^{\tau_2} \mathbf{1}\left(m_\theta(t, s, \mathbf{X}(t)) - c \geq \Sigma_\theta(t, s)y\right) \Sigma_\theta(t, s) \, ds \, dy,$$

*for* $t \leq \tau_1 < \tau_2$ *and* $Z$ *being a standard normally distributed random variable.*

**Proof.**     Using (5.23) in the expression of $F_{\mathrm{CDD}}(t, \tau_1, \tau_2)$ in Prop. 5.4, we derive the Corollary by applying the Fubini-Tonelli Theorem.     □

The dynamics of a CDD futures can be derived by using Itô's Formula, and the result is given in the next Proposition.

**Proposition 5.5.** *The $Q^\theta$-dynamics of $F_{CDD}(t, \tau_1, \tau_2)$ is given as*

$$dF_{\mathrm{CDD}}(t, \tau_1, \tau_2)$$
$$= \sigma(t) \int_{\tau_1}^{\tau_2} \left( \mathbf{b}' \exp(A_\theta(s - t))\mathbf{e}_p \right) \Phi \left( \frac{m_\theta(t, s, \mathbf{X}(t)) - c}{\Sigma_\theta(t, s)} \right) \, ds \, d\widetilde{B}(t),$$

*with $\Phi$ being the cumulative standard normal distribution function.*

**Proof.**     From the knowledge that $F_{\mathrm{CDD}}(t, \tau_1, \tau_2)$ is a $Q^\theta$-martingale, it is only the $d\widetilde{B}$ term appearing in $\mathbf{X}(t)$ that will contribute to the dynamics. Hence, by using the Itô Formula on the expression in Prop. 5.4, we find

$$dF_{\mathrm{CDD}}(t, \tau_1, \tau_2) = d \int_{\tau_1}^{\tau_2} \Sigma_\theta(t, s)\Psi \left( \frac{m_\theta(t, s, \mathbf{X}(t)) - c}{\Sigma_\theta(t, s)} \right) \, ds$$
$$= \int_{\tau_1}^{\tau_2} \Sigma_\theta(t, s)\Psi' \left( \frac{m_\theta(t, s, \mathbf{X}(t)) - c}{\Sigma_\theta(t, s)} \right) \Sigma_\theta^{-1}(t, s)$$
$$\times \mathbf{b}' \exp(A_\theta(s - t))\mathbf{e}_p \sigma(t) \, d\widetilde{B}(t) \, ds.$$

Since $\Psi'(x) = \Phi(x)$, the Proposition follows from the stochastic Fubini Theorem (see [Protter (1990)]).     □

Note that the volatility of the CDD futures becomes stochastically dependent on $\mathbf{X}(t)$, the states in the temperature CARMA model. This is in contrast to the CAT futures dynamics, which has a deterministic volatility (recall Prop. 5.3). The term $\mathbf{b}' \exp(A_\theta(s - t))\mathbf{e}_p$ is scaled by $\Phi((m_\theta(t, s, \mathbf{X}(t)) - c)/\Sigma_\theta(t, s))$ in the volatility of the CDD futures dynamics, whereas for the CAT futures this term is scaled by one.

We move on to HDD futures prices, and present the price and its dynamics in the proposition.

**Proposition 5.6.** *The HDD futures price for $t \le \tau_1 < \tau_2$ is*

$$F_{HDD}(t, \tau_1, \tau_2) = \int_{\tau_1}^{\tau_2} \Sigma_\theta(t, s)\Psi \left( \frac{c - m_\theta(t, s, \mathbf{X}(t))}{\Sigma_\theta(t, s)} \right) \, ds$$

*where $\Psi$ is defined in (5.22) and the dynamics is*

$$dF_{HDD}(t, \tau_1, \tau_2)$$
$$= -\sigma(t) \int_{\tau_1}^{\tau_2} \left( \mathbf{b}' \exp(A_\theta(s - t))\mathbf{e}_p \right) \Phi \left( \frac{c - m_\theta(t, s, \mathbf{X}(t))}{\Sigma_\theta(t, s)} \right) \, ds \, d\widetilde{B}(t).$$

**Proof.** The HDD futures price can be either calculated directly as in the proof of Prop. 5.4, or by applying the HDD-CDD parity in Corollary. 5.1 together with the fact that

$$\Psi(-x) = \Psi(x) - x.$$

The dynamics is derived by the multidimensional Itô Formula as in Prop. 5.5.                                                                      □

Notice that the dynamics of the HDD futures has a minus in front. This signifies that a movement of the CDD futures resulting from $d\widetilde{B}(t) > 0$ gives a movement in the opposite direction for the HDD futures price (resulting from $-d\widetilde{B}(t)$). This is natural in view of the CDD-HDD index parity.

## 5.4   Analysis of temperature futures prices

In this section we analyze the properties of the temperature futures prices, and contrast some of these with empirical findings.

### 5.4.1   *Temperature futures prices and the states of temperature*

We observe from the results in Props. 5.2, 5.4 and 5.6 that the temperature futures prices at time $t$ on the CAT, CDD and HDD indices are all explicitly dependent on the state vector $\mathbf{X}(t)$, and not on the current temperature $T(t)$. This complicates the calibration of the market price of risk parameters $\theta_0$ and $\theta$.

To calibrate these to the market, a natural approach would be to minimize the distance between theoretical and observed futures prices, $F_{\text{Ind}}(t_i, \tau_1, \tau_2; \theta_0, \theta)$ and $\widehat{F}_{\text{Ind}}(t_i, \tau_1, \tau_2)$, respectively, over some set of parameters $(\theta_0, \theta) \in \Theta$. Here we have included $\theta_0$ and $\theta$ in the notation for temperature futures prices to emphasize the dependency on these. In mathematical terms, we want to find $(\theta_0^*, \theta^*) \in \Theta$ minimizing

$$\min_{(\theta_0, \theta) \in \Theta} \left\| F_{\text{Ind}}(\cdot, \tau_1, \tau_2; \theta_0, \theta) - \widehat{F}_{\text{Ind}}(\cdot, \tau_1, \tau_2) \right\|, \qquad (5.24)$$

where $\| \cdot \|$ is some norm measuring the distance between the observed and theoretical prices over the observation times. But for a given choice of $\theta_0$ and $\theta$, we can only derive $F_{\text{Ind}}(t, \tau_1, \tau_2; \theta_0, \theta)$ if we know $\mathbf{X}(t)$ at the observation time $t$. Of course, we have $T(t)$ available in the observation times, and from these we recover $\mathbf{X}(t)$. This can be done by applying one of the filtering methods proposed in Sects. 4.4 and 4.5.

We remark that the minimization in (5.24) can also take different temperature indices and different measurement periods into account. To obtain a consistent choice of $Q^\theta$, one should in fact do that. A specific choice of the norm could simply be the distance measure, possibly weighted to put emphasis on certain observations, for example, more liquid prices compared to less liquid ones.

Let us focus on the situation where we assume $\theta = \mathbf{0}$, that is, we consider pricing measures $Q^\theta$ where the matrix $A$ in the CARMA dynamics $\mathbf{X}(t)$ is unchanged. For this case, the estimation of $\theta_0$ becomes particularly simple, as we can use futures prices sufficiently far from measurement to recover the risk parameter. To simplify our discussion, we focus on CAT futures where we assume $\theta_0(t)$ to be constant, denoted by $\theta_0$.

We recall the CAT futures price from Prop. 5.2 to be

$$F_{\text{CAT}}(t, \tau_1, \tau_2) = \int_{\tau_1}^{\tau_2} \Lambda(s) \, ds + \mathbf{b}' C(\tau_2 - t, \tau_1 - t) \mathbf{X}(t)$$
$$+ \theta_0 \int_t^{\tau_2} C(\tau_2 - s, \tau_1 - s) \mathbf{e}_p \, ds \,,$$

where

$$C(u, v) = A^{-1} \left( \exp(Au) - \exp(Av) \right) \,,$$

for $u \geq v \geq 0$, and

$$C(u, v) = A^{-1} \left( \exp(Au - I) \right) \,,$$

for $v < 0$ and $u \geq 0$. Integrating, we find

$$\int_t^{\tau_2} \mathbf{b}' C(\tau_2 - s, \tau_1 - s) \mathbf{e}_p \, ds$$

$$= \int_t^{\tau_1} \mathbf{b}' C(\tau_2 - s, \tau_1 - s) \mathbf{e}_p \, ds + \int_{\tau_1}^{\tau_2} \mathbf{b}' C(\tau_2 - s, \tau_1 - s) \mathbf{e}_p \, ds$$

$$= \mathbf{b}' A^{-1} \int_t^{\tau_1} \left( e^{A(\tau_2 - s)} - e^{A(\tau_1 - s)} \right) ds \mathbf{e}_p$$

$$+ \mathbf{b}' A^{-1} \int_{\tau_1}^{\tau_2} \left( e^{A(\tau_2 - s)} - I \right) ds \mathbf{e}_p$$

$$= \mathbf{b}' A^{-2} \left( e^{A(\tau_2 - t)} - e^{A(\tau_1 - t)} \right) \mathbf{e}_p - \mathbf{b}' A^{-1} \mathbf{e}_p (\tau_2 - \tau_1) \,.$$

From the assumption of stationarity, the real parts of the eigenvalues of $A$ are negative. Hence, from the spectral representation, terms of $\exp(\tau_i - t)$,

$i = 1, 2$, will tend to zero as $\tau_1 - t \to \infty$ (and then, of course, $\tau_2 - t \to \infty$). This implies in turn that

$$\lim_{\tau_1 - t \to \infty} \left( F_{\text{CAT}}(t, \tau_1, \tau_2) - \int_{\tau_1}^{\tau_2} \Lambda(s)\, ds + \theta_0 \mathbf{b}' A^{-1} \mathbf{e}_p(\tau_2 - \tau_1) \right) = 0.$$

We can use this to estimate the market price of risk $\theta_0$. For example, in the simple case of one CAT futures contract with measurement period $[\tau_1, \tau_2]$, we collect the prices at times $t$ sufficiently far from $\tau_1$, and subtract the aggregated seasonal mean $\Lambda(s)$ over the measurement period from these. We let $\theta_0$ be the constant which minimizes the difference between the deasonalized CAT futures prices and $-\theta_0 \mathbf{b}' A^{-1} \mathbf{e}_p(\tau_2 - \tau_1)$.

As there are no weather derivatives written on locations in Lithuania, we apply price data from Stockholm, Sweden in an empirical example. In [Benth, Šaltytė Benth and Koekebakker (2008)], Chapter 10, we fitted a CAR(3) dynamics to a series of temperatures observed in Stockholm from 1 January 1961 to 25 May 2006. The seasonal function was estimated to be

$$\Lambda(t) = 6.3750 + 0.0001t + 10.4411 \cos(2\pi(t + 165.7591)/365)$$

while the $A$ matrix had parameters

$$A = \begin{bmatrix} 0 & 1 & 0 \\ 0 & 0 & 1 \\ -0.177 & -1.339 & -2.043 \end{bmatrix}.$$

As we estimated the dynamics to be CAR(3), the vector $\mathbf{b}$ becomes $\mathbf{b}' = (1, 0, 0)$. We had accessible daily CAT futures prices from the CME written on Stockholm temperatures measured over the month of July in 2006. The price data were daily settlement prices collected on weekdays from June 5 till June 30, all constantly equal to 480 (this is the futures price after dividing out the money factor 20, see Chapter 1). We aggregate seasonal function over July and find the figure 525.0688. Hence, we have that

$$\widehat{F}_{\text{CAT}}(t, \text{July}) - \sum_{s=1\ \text{July}}^{31\ \text{July}} \Lambda(s) = -45.0688.$$

According to our approach, the market price of risk $\theta_0$ is,

$$\theta_0 = \frac{45.0688}{\mathbf{e}_1 A^{-1} \mathbf{e}_3 (\tau_2 - \tau_1)} = \frac{45.0688}{-5.6497 \times 31} = -0.2573.$$

Thus, the market price of risk is negative, meaning that the market puts a negative drift into the CAR(3) dynamics of Stockholm temperatures. Note that the expected CAT index value for Stockholm in July 2006 would be simply the aggregated seasonal function over July when we are far from measurement period. The futures price is lower than this, which would be in correspondence with the theory for so-called *normal backwardation*, that is, futures prices are lower than the predicted temperature index due to hedging pressure from those who wants to lock in future temperatures. We will come back to this in the next Subsection. For comparison, we had also CAT futures prices for July 2007, being constantly equal to 487. This implies a market price of risk equal to $-0.2238$. We remark that the liquidy in these futures are rather low. Any buyer of these futures will charge a large premium, that is paid by the hedger. The buyer will receive the CAT index, which in expectation will be equal to the aggregated seasonality function over July when viewed far from the measurement period. In return, the buyer pays the CAT futures price. The difference will constitute the "insurance premium" for the hedger to pay.

If we have a collection of CAT futures prices with various measurement periods, we can view the estimation of $\theta_0$ as a linear regression between deseasonalized CAT futures prices and length of measurement period $(\tau_2 - \tau_1)$, that is,

$$\widehat{F}_{\mathrm{CAT}}(t, \tau_1, \tau_2) - \int_{\tau_1}^{\tau_2} \Lambda(s)\, ds = -\theta_0 \mathbf{b}' A^{-1} \mathbf{e}_p (\tau_2 - \tau_1),$$

with $\widehat{F}_{\mathrm{CAT}}$ denoting the observed prices. In [Härdle and Lopez Cabrera (2012)] an extensive empirical study of the market price of risk based on German temperature futures data can be found. The study involves an analysis of different time-dependent specifications of $\theta_0$ (see also [Benth, Härdle and Lopez Cabrera (2011)]).

Remark that the above considerations on the asymptotic behaviour of the CAT futures prices are valid for $\theta \neq \mathbf{0}$ as well whenever $A_\theta$ is stationary. Thus, the market price of risk $\theta$ is only affecting the CAT futures prices close to the measurement period. The asymptotic analysis of CDD futures is similar, but slightly more involved. First, recall the CDD futures price from Prop. 5.4 to be

$$F_{\mathrm{CDD}}(t, \tau_1, \tau_2) = \int_{\tau_1}^{\tau_2} \Sigma_\theta(t, s) \Psi\left(\frac{m_\theta(t, s, \mathbf{X}(t)) - c}{\Sigma_\theta(t, s)}\right) ds.$$

As we have $s \in [\tau_1, \tau_2)$, $s - t \to \infty$ whenever $\tau_1 - t \to \infty$. From Prop. 4.1 it follows that $\Sigma_\theta(t, s)$ converges to some constant $\Sigma_\theta(\infty)$ when $A_\theta$ is a stationary matrix. Furthermore, similarly as in the CAT futures case,

$$\lim_{s-t \to \infty} \left( m_\theta(t, s, \mathbf{X}(t)) - \Lambda(s) + \theta_0 \mathbf{b}' A_\theta^{-1} \mathbf{e}_p \right) = 0.$$

Therefore, after moving the limit inside the $ds$ integral which is permitted by the Fubini-Tonelli theorem (see [Folland (1984)]), we have asymptotically that

$$F_{\text{CDD}}(t, \tau_1, \tau_2) \sim \Sigma_\theta(\infty) \int_{\tau_1}^{\tau_2} \Psi\left(\frac{\Lambda(s) - \theta_0 \mathbf{b}' A_\theta^{-1} \mathbf{e}_p - c}{\Sigma_\theta(\infty)}\right) ds,$$

when $\tau_1 - t \to \infty$. Also in this case we find that the futures prices are constant far from the start of the measurement period. We can utilize this in the calibration of $\theta_0$.

### 5.4.2   *The theoretical risk premium of temperature*

We study here the *risk premium* for the temperature futures market. The risk premium is defined as the difference between the forward price and the predicted temperature index, that is,

$$R_{\text{Ind}}(t, \tau_1, \tau_2) = F_{\text{Ind}}(t, \tau_1, \tau_2) - \mathbb{E}\left[\text{Ind}(\tau_1, \tau_2) \,|\, \mathcal{F}_t\right], \qquad (5.25)$$

where we recall Ind=CAT, CDD or HDD measured over $[\tau_1, \tau_2]$. The expected value of the index $\text{Ind}(\tau_1, \tau_2)$ given current information $\mathcal{F}_t$ is the best prediction at time $t$ of the income from being long a temperature futures on Ind. The market agrees on a price $F_{\text{Ind}}(t, \tau_1, \tau_2)$, which might differ from this prediction. If it is lower, then the risk premium will be negative, which means that the seller of the contract is willing to get less than the best prediction. Compared to commodity markets, this can be viewed as accepting to pay a premium for the insurance of fixing the temperature index rather than being exposed to a floating index. For example, producers of electricity may be interested in securing their production of electricity towards temperature risk, and would like to lock in a certain temperature. The insurance premium could be viewed as the risk premium in this case. This situation corresponds to the futures market being in *normal backwardation*, as the case in the empirical study of CAT prices in Stockholm in the previous Subsection.

Based on our model for temperatures and the analytical expressions for forward prices, we would like to investigate from a theoretical point of view the risk premium. Of course, the premium will be a result of the choice of $\theta_0$ and $\theta$ in the pricing measure $Q^\theta$, and our aim below is to understand this in more detail. Due to analytical simplicity, we shall be concerned with the risk premium in the case of CAT futures. The following holds true.

**Proposition 5.7.** *The risk premium for a CAT futures with price* $F_{CAT}(t, \tau_1, \tau_2)$ *as given in Prop. 5.2 is*

$$R_{CAT}(t, \tau_1, \tau_2) = \mathbf{b}' \left( C_\theta(\tau_2 - t, \tau_1 - t) - C_0(\tau_2 - t, \tau_1 - t) \right) \mathbf{X}(t)$$

$$+ \int_t^{\tau_2} \mathbf{b}' C_\theta(\tau_2 - s, \tau_1 - s) \mathbf{e}_p \theta_0(s) \, ds \,,$$

*where* $C_\theta$ *is defined in* (5.20) *and* (5.21) *and* $A_0 = A$ *in the case of* $C_0$.

**Proof.** The computation of $\mathbb{E}[\int_{\tau_1}^{\tau_2} T(s) \, ds \,|\, \mathcal{F}_t]$ goes exactly as for $F_{CAT}(t, \tau_1, \tau_2)$ in Prop. 5.2 by using $\theta_0 = 0$ and $\theta = \mathbf{0}$, that is, $Q = P$. □

Let us discuss the case $p = 1$ when temperature follows an Ornstein-Uhlenbeck process. To understand the effect of changing the AR parameters, we focus first on the situation where the level is not altered, that is $\theta_0 = 0$. By using the definition of $C_\theta$ in (5.20) and (5.21), the risk premium becomes (recall $\theta < \alpha$ to preserve the AR structure)

$$R_{CAT}(t, \tau_1, \tau_2) = \left( \frac{1}{\alpha - \theta} \left( e^{-(\alpha - \theta)(\tau_1 - t)} - e^{-(\alpha - \theta)(\tau_2 - t)} \right) \right.$$

$$\left. - \frac{1}{\alpha} \left( e^{-\alpha(\tau_1 - t)} - e^{-\alpha(\tau_2 - t)} \right) \right) (T(t) - \Lambda(t))$$

$$= e^{-\alpha(\tau_1 - t)} \rho(\tau_1 - t, \tau_2 - \tau_1) (T(t) - \Lambda(t)) \,,$$

with

$$\rho(x, y) = e^{\theta x} \frac{1 - e^{-(\alpha - \theta)y}}{\alpha - \theta} - \frac{1 - e^{-\alpha y}}{\alpha}. \tag{5.26}$$

Note that in the function $\rho$, $x$ is *time to measurement*, while $y$ is *length of measurement*. To assess the sign of $\rho$, fix an arbitrary $x \geq 0$, and consider the function $y \mapsto \widehat{\rho}(y) := \rho(x, y)$. We have that $\widehat{\rho}(0) = 0$ and

$$\widehat{\rho}'(y) = e^{-\alpha y} \left( e^{\theta(x + y)} - 1 \right).$$

Thus, if $\theta \in (0, \alpha)$, then $\widehat{\rho}'(y) \geq 0$, and therefore $\widehat{\rho}(y) \geq 0$. Otherwise, $\theta < 0$ implies $\widehat{\rho}'(y) \leq 0$, and therefore $\widehat{\rho}(y) \leq 0$. Since $x$ is arbitrary, we conclude that $\rho(x, y)$ is positive whenever $\theta \in (0, \alpha)$ and negative for $\theta < 0$. Of course, $\theta = 0$ implies a zero risk premium, and trivially we find $\rho(x, y) = 0$ confirming this.

We know that the stationary distribution of $X(t)$ under $Q^\theta$ is normal with zero mean and variance equal to $\sigma^2 / (2(\alpha - \theta))$. This means that if we choose $\theta < 0$, then $\alpha - \theta > \alpha$, and hence the variance of $X(t)$ in stationarity is smaller under the risk neutral measure than under the market probability $P$. Hence, choosing $\theta < 0$ reduces the risk in the model. If $\theta \in (0, \alpha)$, then $\alpha - \theta < \alpha$, and thus the risk in the deseasonalized temperatures is larger under the pricing probability compared to $P$. For this case, we obtain a *positive* risk premium whenever $T(t) > \Lambda(t)$, and a negative otherwise. This means that the size and sign of the risk premium are both stochastically varying with the deseasonalized temperatures, and higher than normal temperatures imply a positive premium. But we also see that when $\tau_1 - t \to \infty$, then $\exp(-\alpha(\tau_1 - t))\rho(\tau_1 - t, \tau_2 - \tau_1)$ converges to zero, and the risk premium will then converge to zero as well. Hence, a positive premium is most pronounced for contracts with comparably small time to measurement, and vanishing for the far end of the futures market (that is, for contracts with long time to start of measurement period).

We now look at the case of $\theta_0 \neq 0$. For simplicity, we let $\theta_0$ be a constant. Since $C_\theta$ is a positive function, the additional non-stochastic term in the risk premium

$$\theta_0 \int_t^{\tau_2} C_\theta(\tau_2 - s, \tau_1 - s) \, ds$$

will have the same sign as $\theta_0$. Thus, choosing $\theta_0$, for example, negative, we get into a situation where the sign of the risk premium changes from positive for contracts close to measurement, to negative for contracts which are farther from start of measurement. In fact, we may even have a situation where the risk premium is overall negative when $T(t) < \Lambda(t)$, but has a sign change depending on $\tau_1 - t$ and $\tau_2 - \tau_1$ when $T(t) \geq \Lambda(t)$.

Let us remark that if $\theta = 0$, then we can only have a positive or negative risk premium as long as $\theta_0$ is constant. We may introduce a sign change, but this can only happen if $\theta_0$ is time-dependent, leading to deterministic times for the sign change. Consequently, we see the flexibility introduced in describing various shapes of the risk premium by the class of structure preserving measure changes that we presented in Sect. 5.2.

To be even more specific on the possible shapes of the risk premium, we consider a numerical example where $p = 1$ and constant $\theta_0$. By direct integration, we find the explicit representation of the risk premium

$$R_{\mathrm{CAT}}(t, \tau_1, \tau_2) = e^{-\alpha(\tau_1 - t)}\rho(\tau_t - t, \tau_2 - \tau_1)(T(t) - \Lambda(t)) + \frac{\tau_2 - \tau_1}{\alpha - \theta}\theta_0$$

$$- \frac{\theta_0}{(\alpha - \theta)^2}e^{-(\alpha - \theta)(\tau_1 - t)}\left(1 - e^{-(\alpha - \theta)(\tau_2 - \tau_1)}\right).$$

Let us consider $\alpha = 0.2$ and focus on monthly CAT contracts, where $\tau_2 - \tau_1 = 31$ days. For illustration, we choose $\theta_0 = -0.04$, which gives a negative risk premium far from start of the measurement period. Furthermore, we set $\theta = 0.1$, giving a positive contribution to the risk premium in the "short end" as long as today's temperature is above the seasonal mean. In Fig. 5.1 we draw the implied risk premium as a function of $\tau_1 - t$ for $T(t) - \Lambda(t) = 3$ (solid line) and $T(t) - \Lambda(t) = -3$ (dotted line). As we see, the risk premium

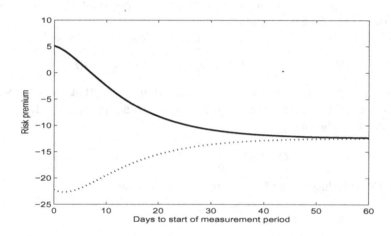

Fig. 5.1   The risk premium as a function of time to start of measurement period, when the CAT contract is measured over 31 days, for $T(t) - \Lambda(t) = 3$ (solid line) and $T(t) - \Lambda(t) = -3$ (dotted line).

in both cases converges to a limit of roughly $-12.5$. When we approach start of measurement, the risk premium increases for the case of the current temperature of 3°C above the seasonal mean. Eventually it goes from a negative to a positive premium for about six days before the measurement period starts. For the other case, the risk premium is decreasing, however, with a slight increase close to measurement period. Note that close to measurement period, it is possible to switch from a negative premium of about $-20$ to a positive premium of about 2-3 in short time if there is a rapid temperature increase of about 6°C. Remark that the premium is not "constant" before about 40 days until maturity and farther. Recall the Stockholm CAT futures example above, where prices were constant for all days until measurement period start. This illustrates that the CAT contracts for Stockholm are very illiquid. One would expect some change in prices when approaching the measurement period.

### 5.4.3 *The Samuelson effect*

In commodity markets, the *Samuelson effect* refers to the convergence of the forward price volatility to the volatility of the underlying (the spot volatility). This effect appears mathematically when the spot has a mean-reverting dynamics, and was first described in [Samuelson (1965)]. In a spot dynamics with mean reversion, the implied forward price will have a volatility exponentially increasing towards the spot volatility as time to maturity of the forward approaches zero.

Recall the price dynamics of a CAT futures contract in Prop. 5.3,

$$dF_{\mathrm{CAT}}(t, \tau_1, \tau_2) = \mathbf{b}' C_0(\tau_2 - t, \tau_1 - t) \mathbf{e}_p \sigma(t) \, d\widetilde{B}(t),$$

where we for notational simplicity assumed $\theta = 0$, that is, no risk-adjustment of the matrix $A$'s mean reversion coefficients $\alpha_i$, $i = 1, \ldots, p$. Consider first the case $p = 1$, where

$$C_0(u, v) = \frac{1}{\alpha} \left( e^{-\alpha v} - e^{-\alpha u} \right).$$

Then the volatility of the CAT futures price becomes

$$\Sigma_{\mathrm{CAT}}(t, \tau_1, \tau_2) = \frac{\sigma(t)}{\alpha} \left( e^{-\alpha(\tau_1 - t)} - e^{\alpha(\tau_2 - t)} \right)$$

$$= \frac{\sigma(t)}{\alpha} e^{-\alpha(\tau_1 - t)} \left( 1 - e^{-\alpha(\tau_2 - \tau_1)} \right).$$

As the underlying of the CAT futures is temperature, the "spot volatility" is given as the temperature volatility, namely $\sigma(t)$. As time to maturity $\tau_1 - t \to 0$, the forward volatility will increase exponentially with a limit given by

$$\lim_{\tau_1 - t \to 0} \Sigma_{\mathrm{CAT}}(t, \tau_1, \tau_2) = \frac{\sigma(t)}{\alpha} \left( 1 - e^{-\alpha(\tau_2 - \tau_1)} \right).$$

The limit is *not* equal to the temperature volatility $\sigma(t)$, but is modified by a function of the length of measurement period $\tau_2 - \tau_1$. This is not surprising, in view of the fact that the CAT futures is settled against the aggregated temperature over the measurement period, and not only at the temperature in a specific time point. Denoting $f(x) = (1 - \exp(-\alpha x))/\alpha$, we find that $f(0) = 0$ and that $f(x) \to 1/\alpha$ when $x \to \infty$. Moreover, $f'(x) = \exp(-\alpha x) > 0$ and thus $f(x)$ is increasing. Since $f(x) = 1$ for $x = -(\ln(1 - \alpha))/\alpha$, we find

$$\lim_{\tau_1 - t \to 0} \Sigma_{\mathrm{CAT}}(t, \tau_1, \tau_2) \leq \sigma(t),$$

for $\tau_2 - \tau_1 \leq -(\ln(1-\alpha))/\alpha$, while

$$\lim_{\tau_1 - t \to 0} \Sigma_{\mathrm{CAT}}(t, \tau_1, \tau_2) \geq \sigma(t),$$

otherwise. Note that $\alpha < 1$ in a stationary model, which is the case here. For "short" measurement periods, the CAT futures volatility converges to a value smaller than the temperature volatility as time to measurement period approaches zero, and to a value bigger than $\sigma(t)$ for longer measurement periods. If $\alpha = 0.2$, say, which is not an unreasonable speed of mean reversion for the temperature dynamics, we find that $-(\ln(1-\alpha))/\alpha \approx 1.1$. Hence, if the measurement period is longer than one day, the futures volatility will be above the temperature volatility in the limit. For CAT futures, we hence have a *modified* Samuelson effect as a result of the measurement period.

We consider the case $p > 1$, and focus on the CAR($p$) models where $\mathbf{b} = \mathbf{e}_1$. The Samuelson effect will change character dramatically in this case. The CAT futures price volatility becomes

$$\Sigma_{\mathrm{CAT}}(t, \tau_1, \tau_2) = \sigma(t)\mathbf{e}_1' A^{-1} \left( e^{A(\tau_2 - t)} - e^{A(\tau_1 - t)} \right) \mathbf{e}_p.$$

From the spectral representation of the exponential of $A$, we can express $\Sigma_{\mathrm{CAT}}$ as a sum of exponentially dampened trigonometric functions. Factorizing out $\exp(A(\tau_1 - t))$ gives the expression

$$\Sigma_{\mathrm{CAT}}(t, \tau_1, \tau_2) = \sigma(t)\mathbf{e}_1' A^{-1} e^{A(\tau - 1 - t)} \left( e^{A(\tau_2 - \tau_1)} - I \right) \mathbf{e}_p. \tag{5.27}$$

We use a numerical example to analyze this volatility further as a function of time to measurement $\tau_1 - t$. To this end, let $A$ be given as for Stockholm presented in Subsect. 5.4.1 above, where $p = 3$. Further, we assume $\sigma(t) = 1$ for simplicity, and consider a CAT futures with a monthly measurement period, $\tau_2 - \tau_1 = 31$. In Fig. 5.2 we plot the resulting CAT futures volatility as a function of time to measurement $\tau_1 - t$ (solid line). As expected, the volatility decreases with time to measurement period. But a particular feature for the case $p > 1$ shows up when the contract has short time left until measurement period. It decreases slower. We include the case of a weekly measurement period for comparison (dotted line) to demonstrate the interesting fact that the volatility decreases when time to measurement approaches zero for the last few days. This is a striking fact that violates the classical Samuelson effect, as the volatility first increases, and next decreases with time to measurement.

The explanation for this *modified* Samuelson effect is the AR(3) structure of the underlying temperature dynamics. An AR(3) structure "remembers" three days back, and thus, close to start of measurement period

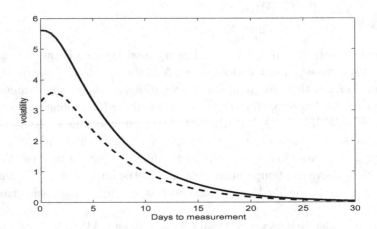

Fig. 5.2   The CAT volatility function in (5.27) for constant $\sigma(t) = 1$, $p = 3$ and $A$ matrix taken from Stockholm data. The solid line depicts a monthly measurement period, while the dotted a weekly.

one can already with small degree of uncertainty predict the temperature *inside* the measurement period. As this is short, the variability will be smaller than farther from the start of measurement. This also explains lower volatility in the case of a weekly measurement period. Something similar happens for a monthly measurement period, but the period is so long that we do not get lower volatility, but only a reduced rate of decrease close to measurement period. It remains to check out if this can be observed in actual CAT futures price data as well.

## 5.5   Pricing wind speed forwards

We recall the Nordix wind speed index discussed in Chapter 1, Subsect. 1.2.2, to be the acummulated wind speed deviation from a historical mean, benchmarked at 100. Let the measurement period for the index be $[\tau_1, \tau_2]$. Then,

$$N(\tau_1, \tau_2) = 100 + \sum_{\tau=\tau_1}^{\tau_2} W(\tau) - w_{20}(\tau)\,.$$

Here, $w_{20}(\tau)$ is the 20 years average wind speed on day $\tau$. We analyze in this Section a continuous analogue of the Nordix index, defined (by a slight abuse of notation) as

$$N(\tau_1, \tau_2) = c(\tau_1, \tau_2) + \int_{\tau_1}^{\tau_2} W(\tau)\, d\tau\,, \tag{5.28}$$

for some function $c$ depending on the measurement period. Note that apart from the function $c$, the wind speed index $N$ coincides with the CAT index for temperatures in the sense that measurements of the underlying weather event are aggregated over a period. If we let

$$c(\tau_1, \tau_2) = 100 - \int_{\tau_1}^{\tau_2} w_{20}(\tau)\, d\tau\,,$$

we obtain the Nordix wind speed index.

Our aim is to derive the forward price of a contract written on the wind speed index. This has been studied in [Benth and Šaltytė Benth (2009)], and we follow the analysis presented there closely in what follows.

Denote by $F_N(t, \tau_1, \tau_2)$ the forward price at time $t$ written on the index $N(\tau_1, \tau_2)$ measured over the time interval $[\tau_1, \tau_2]$, with $t \leq \tau_1$. As for temperature forwards, we define the wind speed forward price as

$$F_N(t, \tau_1, \tau_2) = \mathbb{E}_Q\left[ N(\tau_1, \tau_2)\,|\,\mathcal{F}_t \right]\,, \tag{5.29}$$

for some pricing measure $Q$ being an equivalent probability to $P$. We restrict our considerations to pricing measures given by $Q^\theta$ as defined in Sect. 5.2. Note that we have

$$F_N(t, \tau_1, \tau_2) = c(\tau_1, \tau_2) + \int_{\tau_1}^{\tau_2} \mathbb{E}_Q[W(\tau)\,|\,\mathcal{F}_t]\, d\tau\,. \tag{5.30}$$

We introduce the notation

$$f_N(t, \tau) = \mathbb{E}_Q[W(\tau)\,|\,\mathcal{F}_t] \tag{5.31}$$

as the forward price of a contract "delivering" the wind speed at time $\tau$, $0 \leq t \leq \tau$. Since

$$F_N(t, \tau_1, \tau_2) = c(\tau_1, \tau_2) + \int_{\tau_1}^{\tau_2} f_N(t, \tau)\, d\tau\,, \tag{5.32}$$

it is sufficient to study $f_N(t, \tau)$ in order to find the wind speed forward price. We next study this problem under the assumption that $Q = Q^\theta$. The model for the wind speed $W(\tau)$ is given in Subsect. 4.6.2, where we restrict our attention to Box-Cox transforms with index $\lambda \in [0, 1]$.

For $\lambda \in (0, 1]$ and the definition of the wind speed dynamics, we find as for temperature that conditioned on $\mathcal{F}_t$, it holds

$$\frac{1}{\lambda}(W(\tau)^\lambda - 1) = m_\theta(t, \tau, \mathbf{X}(t)) + Z\Sigma_\theta(t, \tau)\,, \tag{5.33}$$

where $Z$ is a standard normally distributed random variable and equality is in distribution. The mean $m_\theta$ and variance $\Sigma_\theta^2$ are defined in (5.17) and (5.18), respectively. If $\lambda = 0$, we have similarly, that

$$\ln W(\tau) = m_\theta(t, \tau, \mathbf{X}(t)) + Z\Sigma_\theta(t, \tau). \tag{5.34}$$

Before stating the result on wind speed forwards, introduce the notation $M_k(a, b^2)$ as the $k$th moment of a normally distributed random variable with mean $a$ and variance $b^2$, that is

$$M_k(a, b^2) = \frac{1}{\sqrt{2\pi b^2}} \int_{\mathbb{R}} x^k \exp\left(-\frac{(x-a)^2}{2b^2}\right) dx. \tag{5.35}$$

By a simple change of variables, we obtain the equivalent expression

$$M_k(a, b^2) = \frac{1}{\sqrt{2\pi}} \int_{\mathbb{R}} (a + by)^k \exp\left(-\frac{1}{2}y^2\right) dy. \tag{5.36}$$

We have the following proposition.

**Proposition 5.8.** *For $\lambda \in (0, 1]$ and $0 \le t \le \tau$ it holds that*

$$f_N(t, \tau) = M_{\frac{1}{\lambda}}\left(1 + \lambda m_\theta(t, \tau, \mathbf{X}(t)), \lambda^2 \Sigma_\theta^2(t, \tau)\right). \tag{5.37}$$

*When $\lambda = 0$ and $0 \le t \le \tau$ we have*

$$f_N(t, \tau) = \exp\left(m_\theta(t, \tau, \mathbf{X}(t)) + \frac{1}{2}\Sigma_\theta^2(t, \tau)\right). \tag{5.38}$$

***Proof.*** Let $\lambda > 0$. From (5.33) we know that with respect to the pricing measure $Q$, the wind speed at time $\tau \ge t$ conditioned on $\mathcal{F}_t$ for $t \ge 0$ is (with equality in distribution)

$$W(\tau) = \left(\lambda(m_\theta(t, \tau, \mathbf{X}(t)) + Z\Sigma_\theta(t, \tau)) + 1\right)^{1/\lambda},$$

where $Z$ is a standard normally distributed random variable independent of $\mathcal{F}_t$. We observe further that $m_\theta(t, \tau, \mathbf{X}(t))$ is $\mathcal{F}_t$-measurable. A direct calculation gives the result. The case $\lambda = 0$ follows similarly.     $\square$

The forward on the Nordix wind speed index will now become the integral of the function $\tau \mapsto f_N(t, \tau)$ over the measurement period $[\tau_1, \tau_2]$, adding the level $c(\tau_1, \tau_2)$. As for the case of temperature forwards, the wind speed forward price $F_N(t, \tau_1, \tau_2)$ *does not* depend on the current wind speed $W(t)$, but explicitly on the *factors* $\mathbf{X}(t)$ in the CARMA model.

From Itô's Formula, we find the dynamics of $f_N(t, \tau)$ in the next Proposition.

**Proposition 5.9.** *For* $0 \le t \le \tau$ *we have*

$$\frac{df_N(t,\tau)}{f_N(t,\tau)} = g_\lambda(t,\tau,\mathbf{X}(t)) \Big(\mathbf{b}' \exp(A_\theta(\tau - t))\mathbf{e}_p\Big)\sigma(t)\, d\widetilde{B}(t),$$

*where* $g_0(t,\tau,\mathbf{X}(t)) = 1$ *and*

$$g_\lambda(t,\tau,\mathbf{X}(t)) = \frac{M_{\frac{1}{\lambda}-1}\big(1 + \lambda m_\theta(t,\tau,\mathbf{X}(t)), \lambda^2 \Sigma_\theta^2(t,\tau)\big)}{M_{\frac{1}{\lambda}}\big(1 + \lambda m_\theta(t,\tau,\mathbf{X}(t)), \lambda^2 \Sigma_\theta^2(t,\tau)\big)},$$

*for* $\lambda \in (0,1]$.

***Proof.*** First, by elementary computations using (5.36) we get that

$$\frac{\partial}{\partial a} M_k(a,b^2) = k M_{k-1}(a,b^2).$$

Applying the multidimensional Itô Formula in Thm. 4.1, we find from Prop. 5.8 that

$$df(t,\tau) = \frac{\partial}{\partial a} M_{\frac{1}{\lambda}}(1 + \lambda m_\theta(t,\tau,\mathbf{X}(t)), \lambda^2 \Sigma_\theta^2(t,\tau))$$
$$\times \lambda \mathbf{b}' \exp(A_\theta(\tau - t))\mathbf{e}_p\sigma(t)\, d\widetilde{B}(t).$$

Thus, by using Prop. 5.8 again, the result follows for $\lambda \in (0,1]$. The case of $\lambda = 0$ follows easily from the multidimensional Itô Formula in Thm 4.1. $\square$

Remark that by using expression in (5.35) for the moment generating function $M_k(a,b^2)$, we find

$$\frac{\partial}{\partial a} M_k(a,b^2) = \frac{1}{b^2} M_{k+1}(a,b^2) - \frac{a}{b^2} M_k(a,b^2).$$

This will lead to a function $g_\lambda$ given as

$$g_\lambda(t,\tau,\mathbf{X}(t)) = \frac{1}{\lambda \Sigma_\theta^2(t,\tau)} \Big(\frac{M_{\frac{1}{\lambda}+1}\big(1 + \lambda m_\theta(t,\tau,\mathbf{X}(t)), \lambda^2 \Sigma_\theta^2(t,\tau)\big)}{M_{\frac{1}{\lambda}}\big(1 + \lambda m_\theta(t,\tau,\mathbf{X}(t)), \lambda^2 \Sigma_\theta^2(t,\tau)\big)}$$
$$- (1 + \lambda m_\theta(t,\tau,\mathbf{X}(t)))\Big),$$

for $\lambda \in (0,1]$. Of course, the two formulas for $g_\lambda$ coincide.

## Chapter 6

# Extensions of temperature and wind speed models

In the previous two Chapters we analyzed stochastic models for the temperature and wind speed dynamics based on CARMA processes. The purpose of this Chapter is to extend these models in various directions, to allow for even more sophisticated dynamics for these weather factors. In particular, we introduce stochastic volatility in weather modelling, and generalize the CARMA dynamics to so-called Lévy semistationary processes. The popular class of fractional Brownian motion based models is discussed. For all the models we look at the pricing of weather derivatives as an application.

In the analyses of this Chapter we make use of Lévy processes. The reader not familiar with this class of stochastic processes, can find a basic, yet thorough introduction in [Cont and Tankov (2004)].

## 6.1 Stochastic temperature volatility

In the study of temperatures in US cities, [Carmona and Diko (2005)] detect clear signs of stochastic volatility. They argue empirically for a seasonal GARCH effects by looking at the squared residuals from an AR model. [Benth and Šaltytė-Benth (2005)] find evidence of stochastic volatility in their study of Norwegian temperature data, while [Benth, Härdle and Lopez Cabrera (2011)] observe stochastic volatility in Asian temperatures. We also refer to the statistical analysis of Lithuanian temperature and wind speed data in Chapter 3.

[Carmona and Diko (2005)] suggest to model the stochastic volatility of the temperature dynamics by a sum of a seasonal function and a GARCH process. The unfortunate effect of such a specification is that one can obtain negative values of the volatility, although with small probability. In [Benth and Šaltytė Benth (2011)] a multiplicative model for the volatility is proposed as an attractive alternative, where a stepwise procedure to estimate the model can be used. We next present this model, which also has obvious applications to wind speed modelling.

Suppose $\mathbf{X}(t)$ follows the dynamics

$$d\mathbf{X}(t) = A\mathbf{X}(t)\,dt + \phi(t)\mathbf{e}_p\,dB(t)\,, \tag{6.1}$$

where

$$\phi(t) = \sigma(t)\sqrt{V(t)}\,, \tag{6.2}$$

for a bounded positive continuous function $\sigma(t)$, and $V(t)$ being a positive stochastic process. Here, we assume that $\sigma(t)$ accounts for the seasonal variations of the volatility, and is of the form that we observed in the empirical studies in Chapters 3 and 4, and later in Chapter 5 when pricing derivatives.

A class of stochastic volatility processes providing a great deal of flexibility in precise modelling of residual characteristics is given by the Barndorff-Nielsen and Shephard (BNS) model (see [Barndorff-Nielsen and Shephard (2001)]). In mathematical terms, we have

$$dV(t) = -\lambda V(t)\,dt + dL(t)\,. \tag{6.3}$$

Here, $\lambda > 0$ is a constant measuring the speed of mean reversion for the volatility process $V(t)$, which reverts to zero. The process $L(t)$ is assumed to be a *subordinator* independent of $B$, the Brownian motion, meaning a Lévy process with increasing paths. A Lévy process generalizes Brownian motion because it is a stochastic process with independent and stationary increments, but where the increments are not necessarily normally distributed. We refer to [Cont and Tankov (2004)] for an introduction to this class of stochastic processes, popular in modelling financial markets. Remark that the BNS stochastic volatility model does not become a GARCH dynamics in a discrete-time setting, but shares some similar properties. Its analytical tractability is very advantageous when applying it in practice. A detailed analysis of the various aspects discussed below can be found in [Barndorff-Nielsen and Shephard (2001)].

Restricting to Lévy processes $L(t)$ with only positive jumps, ensures that $V(t)$ remains positive. By the Itô Formula for jump processes (see [Ikeda and Watanabe (1981)]), it is possible to show that

$$V(t) = V(0)e^{-\lambda t} + \int_0^t e^{-\lambda(t-s)}\,dL(s)\,. \tag{6.4}$$

We compute the covariance between $V(t)$ and $V(t+k)$ for $k > 0$ by appealing to the independent increment property of the Lévy process $L$ and the isometry for integrals:

$$\text{Cov}\,(V(t+k), V(t))$$

$$= \text{Cov}\left(\int_0^{t+k} e^{-\lambda(t+k-s)}\,dL(s), \int_0^t e^{-\lambda(t-s)}\,dL(s)\right)$$

$$= \mathbb{E}\left[\int_0^{t+k} e^{-\lambda(t+k-s)}\,dL(s)\int_0^t e^{-\lambda(t-s)}\,dL(s)\right]$$

$$- \mathbb{E}\left[\int_0^{t+k} e^{\lambda(t+k-s)}\,dL(s)\right]\mathbb{E}\left[\int_0^t e^{-\lambda(t-s)}\,dL(s)\right]$$

$$= e^{-\lambda k}\text{Var}\left(\int_0^t e^{-\lambda(t-s)}\,dL(s)\right)$$

$$= \frac{\text{Var}(L(1))}{2\lambda}(1 - e^{-2\lambda t}) \times e^{-\lambda k}\,.$$

Thus, letting $t \to \infty$ we find the *stationary* ACF of $V(t)$ to be

$$\rho(k) = \lim_{t\to\infty} \text{Corr}(V(t+k), V(t)) = \exp(-\lambda k)\,. \tag{6.5}$$

Hence, the BNS model yields an exponentially decaying ACF for the stochastic volatility.

Let the residuals for a time discretization $\Delta > 0$ be

$$R(t) := \sqrt{V(t)}\Delta B(t)\,, \tag{6.6}$$

with $\Delta B(t) = B(t+\Delta) - B(t)$. As long as $k \geq \Delta$, we find that

$$\mathbb{E}[R^2(t+k)R^2(t)] = \mathbb{E}[V(t+k)V(t)]\Delta^2$$

by the independent increment property of Brownian motion and the $\mathcal{F}_t$ adaptedness of $V(t)$. Hence,

$$\text{Corr}(R^2(t+k), R^2(t)) = \text{Corr}(V(t+k), V(t)) = \rho(k)\,.$$

Therefore, if we have available residual data $R(t)$, we can find $\lambda$, the speed of mean reversion of the volatility by simply fitting an exponential function to the empirical ACF of the squared residuals. In practice, it may turn out that such an exponential function is too simple to model the empirical ACF. In that case, we can extend the above stochastic volatility model to be a superposition of processes of the type $V(t)$. This would yield a theoretical ACF being a sum of exponentially decaying functions with different rates.

Such an extension provides a great deal of flexibility in modelling multi-scale mean reversion in the stochastic volatility.

Following the properties of Lévy processes, one can explicitly compute the (logarithm of the) moment generating function of $V(t)$ to be

$$\ln \mathbb{E}\left[\exp(xV(t))\right] = xV(0)\exp(-\lambda t) + \int_0^t \phi(x\exp(-\lambda s))\,ds\,,$$

where $\phi(x)$ is the logarithm of the moment generating function of $L(1)$. Note in passing that the moment generating function of $L(1)$ characterizes the distribution of the increments of the Lévy process $L$, as any Lévy process has stationary and independent increments. This is equivalent to the distribution of $L(t) - L(s)$ for any pair $t > s$ being *infinitely divisible*, and completely characterized by the distribution of $L(1)$. For more on infinitely divisible distributions and Lévy processes, we refer the reader to the monograph by [Sato (1999)].

Letting time $t \to \infty$, we find under mild conditions on $\phi$ that there exists a stationary distribution with moment generating function characterized by

$$\lim_{t\to\infty} \ln \mathbb{E}\left[\exp(xV(t))\right] = \int_0^\infty \phi(x\exp(-\lambda s))\,ds\,.$$

This is the stationary distribution of $V(t)$. In these considerations, it is necessary to assume that $L$ has exponential moments. By conditioning on $V(t)$, we see that

$$\sqrt{V(t)}\,dB(t)|_{V(t)} \sim \mathcal{N}(0, V(t))\,,$$

that is, centered normally distributed with variance $V(t)$. Hence, the residuals become Gaussian variance-mixture models.

One of the ideas in the BNS model is to separate the pathwise dependency structure and the stationary distribution of $V$. The pathwise dependency structure is described by $\lambda$ in the ACF, while knowing the stationary distribution specifies indirectly the distribution of the Lévy process increments. It is convenient not to have a dependency on $\lambda$ in the stationary distribution, and to achieve this one can define $L(t) := U(\lambda t)$, for a subordinator $U$. Hence, we simply scale time by the speed of mean reversion. From the stationarity of Lévy processes, we find that the logarithmic moment generating function of $L$, $\phi_L$, is given by

$$\phi_L(x) = \lambda \phi_U(x)\,,$$

where $\phi_U$ is the logarithmic moment generating function of $U$. Thus, the stationary distribution of $V(t)$ has a moment generating function given by

$$\lim_{t\to\infty} \ln \mathbb{E}[\exp(xV(t))] = \int_0^\infty \phi_U(x\exp(-s))\,ds\,.$$

We see that the stationary distribution of $V(t)$ is independent of $\lambda$. Hence, we can fit a distribution to the residuals independently of the estimate of $\lambda$.

Recall Fig. 3.9 in Chapter 3 on temperature modelling, where the ACF of the squared residuals in Vilnius obtained after explaining the AR structure of deseasonalized DAT and the seasonality in the daily variances is plotted. The seasonality of variances is modelled by $\sigma(t)$ using a truncated Fourier series, as we recall from the empirical analysis in Chapter 3. The ACF decays for the first few lags rapidly down towards zero, and becomes insignificant. The positive correlation structure for small lags points towards stochastic volatility, which can be modelled by the approach suggested above. If an exponentially decaying function is fitted to this empirical ACF, the decay rate as the estimate of $\lambda$ can be used. In [Benth and Šaltytė Benth (2011)] we performed such a study in more detail, where the specification of the subordinator is also discussed in view of temperatures.

Appealing to the Itô Formula for jump processes (see [Ikeda and Watanabe (1981)]) again, we can state the explicit dynamics of $\mathbf{X}(t)$ given $\mathbf{X}(s)$, for $t \geq s$, as

$$\mathbf{X}(t) = \exp(A(t-s))\mathbf{X}(s) + \int_s^t \exp(A(t-u))\mathbf{e}_p\sigma(u)\sqrt{V(u)}\,dB(u)\,. \quad (6.7)$$

From this we may define a CARMA process with *seasonal stochastic volatility* as

$$Y(t) = \mathbf{b}'\mathbf{X}(t)\,. \quad (6.8)$$

Here, $\mathbf{b}$ is the $p$-dimensional vector as defined in Sect. 4.1. This class of models can be embedded into so-called Brownian semistationary processes which we discuss next.

## 6.2 Brownian semistationary processes

A natural generalization of the stationary CARMA processes studied and analyzed in this book is the so-called Brownian semistationary processes, or BSS processes for short. These processes were first introduced in turbulence modelling by [Barndorff-Nielsen and Schmiegel (2004)], and have later been extended to modelling the dynamics of cancer growth (see [Barndorff-Nielsen and Schmiegel (2007, 2009)]) and prices in energy markets (see [Barndorff-Nielsen, Benth and Veraart (2010, 2012)]). In this Section we discuss these processes as a convenient modelling device for weather, in particular wind speed and temperature.

A BSS process is defined to be

$$X(t) = \int_{-\infty}^{t} g(t - s)\sigma(s) \, dB(s), \qquad (6.9)$$

for a real-valued function $g$ on $[0, \infty)$ and an $\mathcal{F}_t$-adapted process $\sigma(t)$ such that for each $t \geq 0$

$$\mathbb{E}\left[\int_{0}^{\infty} g^2(u)\sigma^2(t - u) \, du\right] < \infty.$$

The process $\sigma(t)$ is assumed to be independent of $B(t)$, interpreted as the stochastic volatility process. From the general theory of BSS processes (see e.g. [Barndorff-Nielsen and Schmiegel (2009)]), it is known that $X(t)$ is a zero-mean stationary process, as long as $\sigma(t)$ is stationary. Since $\sigma(t)$ is independent of $B$, we have that $X(t)$ is a variance-mixture model, with $X(t)$ conditioned on the path of $\sigma$, which is normally distributed with variance $\int_0^\infty g^2(u)\sigma^2(t - u) \, du$. Note the unusual timing in the stochastic integral (6.9) defining the BSS process. By integrating from $-\infty$, and not from the more common time zero, we ensure that $X$ is already specified *in stationarity*. This is convenient from an empirical perspective. To illustrate this, we recall the CARMA dynamics of the temperature and wind speed models in Sect. 4.1. For each observation $\mathbf{X}(t)$, it will take some time before the process becomes stationary. With the specification in (6.9), we consider an observation of the temperature as an observation from the stationary process, and not a hard conditioning of the dynamics.

Let us discuss the case of no volatility, that is, $\sigma(t) = 1$. From the Itô isometry we see that $X(t)$ is a stationary Gaussian process with mean zero and variance

$$\text{Var}(X(t)) = \int_{0}^{\infty} g^2(u) \, du. \qquad (6.10)$$

In general, $X(t)$ is not a semimartingale process. However, if we assume $g(0)$ to be well-defined and differentiable, where $g'(t - s)B(s)$ is Lebesgue integrable on $(-\infty, t]$ and $B(s)g(t - s)$ converges to zero in variance when $s \to -\infty$, by applying the Itô Formula we find that

$$\int_{-\infty}^{t} g(t - s) \, dB(s) = B(t)g(0) + \int_{-\infty}^{t} B(s)g'(t - s) \, ds.$$

Hence, $X(t)$ can be represented as a sum of a martingale and a bounded variation process, which defines a semimartingale.

Let now

$$g(u) = \mathbf{b}' \exp(Au)\mathbf{e}_p,$$

for the matrix $A$ and vector $\mathbf{b}$ used in the definition of CARMA processes in Sect. 4.1. Assuming that $A$ has eigenvalues with negative real parts, the process $X(t)$ will be the stationary representation of the CARMA process $Y(t)$. This shows that the family of CARMA models is a special case in the BSS framework. We may obviously incorporate a deterministic volatility process $\sigma(t)$ into the BSS definition. Furthermore, the CARMA model with seasonal stochastic volatility presented in Sect. 6.1 is also a special case of an BSS process.

Let us analyze the pricing of a weather derivative when using BSS processes as the general modelling class. We use the case of wind speed as the example. To keep matters simple, we assume an exponential dynamics for the wind speed, defined at time $t$ as

$$W(t) = \exp(\Lambda(t) + X(t)). \tag{6.11}$$

Hence, the logarithmic deseasonalized wind speed is described by a BSS process $X(t)$ for some general function $g$ (recall the wind speed dynamics for CARMA models in Sect. 4.6). Assume for further simplicity that the stochastic volatility has no seasonality, i.e. $\sigma(t) = 1$.

From Sect. 5.5, we know that the price of a forward contract at time $t$, delivering the Nordix wind speed index $N(\tau_1, \tau_2)$ over the delivery period $[\tau_1, \tau_2]$ is

$$F_N(t, \tau_1, \tau_2) = c(\tau_1, \tau_2) + \int_{\tau_1}^{\tau_2} f_N(t, \tau) \, d\tau,$$

for some deterministic function $c$ and

$$f_N(t, \tau) = \mathbb{E}_Q[W(\tau) \mid \mathcal{F}_t].$$

As usual, $Q$ is a pricing measure. Consider now the change of measure given by the Girsanov transform

$$d\widetilde{B}(t) = \theta_0 \, dt + dB(t),$$

for a constant $\theta_0$ (see Sect. 5.2, and in particular Prop. 5.1). Then, under the probability $Q$, $\widetilde{B}$ is a Brownian motion, and the BSS process has $Q$-dynamics

$$X(t) = \int_{-\infty}^{t} g(t - s) \, d\widetilde{B}(s) - \theta_0 \int_{0}^{\infty} g(s) \, ds. \tag{6.12}$$

Note here that we have to assume that the last integral is finite since square-integrability does not imply integrability. Thus, we introduce an additional assumption that $\int_0^\infty |g(s)|\, ds < \infty$. We compute the conditional expectation of wind speed at time $\tau$ as

$$\mathbb{E}_Q[\exp(X(\tau)) \,|\, \mathcal{F}_t] = \exp\left(-\theta_0 \int_0^\infty g(s)\, ds\right)$$

$$\times \mathbb{E}_Q\left[\exp\left(\int_{-\infty}^\tau g(\tau - s)\, d\tilde{B}(s)\right)\,\Big|\, \mathcal{F}_t\right]$$

$$= \exp\left(-\theta_0 \int_0^\infty g(s)\, ds + \int_{-\infty}^t g(\tau - s)\, d\tilde{B}(s)\right)$$

$$\times \mathbb{E}_Q\left[\exp\left(\int_t^\tau g(\tau - s)\, d\tilde{B}(s)\right)\,\Big|\, \mathcal{F}_t\right].$$

In the second equality above, the $\mathcal{F}_t$-measurability of the Itô integral was applied. Next, using the independent increment property of Brownian motion, we find, applying the normality of the integral $\int_t^\tau g(\tau - s)\, d\tilde{B}(s)$, that

$$\mathbb{E}_Q[\exp(X(\tau)) \,|\, \mathcal{F}_t]$$

$$= \exp\left(-\theta_0 \int_0^\infty g(s)\, ds + \int_{-\infty}^t g(\tau - s)\, d\tilde{B}(s)\right)$$

$$\times \mathbb{E}_Q\left[\exp\left(\int_t^\tau g(\tau - s)\, d\tilde{B}(s)\right)\right]$$

$$= \exp\left(-\theta_0 \int_0^\infty g(s)\, ds + \frac{1}{2}\int_0^{\tau-t} g^2(s)\, ds + \int_{-\infty}^t g(\tau - s)\, d\tilde{B}(s)\right).$$

Hence, the logarithm of the instant delivery forward price $f_N(t, \tau)$ becomes

$$\ln\left(f_N(t, \tau)\right) = \Lambda(\tau) + \int_{-\infty}^t g(\tau - s)\, d\tilde{B}(s) - \theta_0 \int_0^\infty g(s)\, ds$$

$$+ \frac{1}{2}\int_0^{\tau-t} g^2(s)\, ds. \tag{6.13}$$

Although the wind speed is modelled by a one-factor BSS process $X(t)$, it will for general specifications of $g$ not give a forward price depending on the current wind speed, rather some "perturbation" of it. The problem is the detachement of time in the kernel function and the integration in the forward price, namely the term $\int_{-\infty}^t g(\tau - s)\, d\tilde{B}(s)$. For an exponential specification, like in the Ornstein-Uhlenbeck case, with $g(s) = \exp(-\alpha s)$, we find

$$g(\tau - s) = g(\tau - t)g(t - s),$$

incorporating an explicit dependence on the wind speed in the forward price. For other choices of $g$, such a factorization cannot be expected. In the case of $g$ stemming from a CARMA process, it appears as a dependency on the states $\mathbf{X}(t)$ in the forward price, and not on the current wind speed $W(t)$ (see Prop. 5.8).

An interesting interpretation of this fact is that the forward price $f_N(t, \tau)$ is positively correlated with the current wind speed, but not perfectly. It can be shown by computing the characteristic function that $(\ln(f_N(t, \tau)), \ln(W(t)))$ is a bivariate normal variable, with a correlation given as

$$\text{corr}(\ln(f_N(t, \tau)), \ln(W(t)))$$

$$= \text{corr}\left(\int_{-\infty}^{t} g(\tau - s)\, d\widetilde{B}(s), \int_{-\infty}^{t} g(t - s)\, d\widetilde{B}(s)\right)$$

$$= \frac{\int_0^\infty g(\tau - t + x)g(x)\, dx}{\sqrt{\int_0^\infty g^2(\tau - t + x)\, dx \int_0^\infty g^2(x)\, dx}}.$$

This follows from the Itô isometry. We see that the correlation is stationary in *time to delivery* $\tau - t$, and converges to one as $\tau - t \to 0$. Also, the correlation is positive for all $\tau - t \geq 0$ if $g$ is positive. By the Cauchy-Schwarz inequality (see [Folland (1984)]), we have

$$\int_0^\infty g(\tau - t + x)g(x)\, dx \leq \sqrt{\int_0^\infty g^2(\tau - t + x)\, dx \int_0^\infty g^2(x)\, dx},$$

and therefore the correlation is less than one for $\tau - t > 0$. In fact, for most choices of $g$ the Cauchy-Schwarz inequality will be strict when $\tau - t > 0$, yielding a correlation strictly less than one for time to deliveries away from zero. Hence, in the case of BSS modelling of wind speed, we may interpret the relationship between wind speed and forward prices as a kind of linear regression (on a logarithmic scale) rather than an explicit relationship. This opens up for a greater flexibility in the joint modelling of weather factors and forward prices.

## 6.3 Fractional models

From an empirical analysis of temperatures collected in London in the period 1772 to 1992, [Brody, Syroka and Zervos (2002)] find empirical evidence for long-range temporal dependencies in the data. Moreover, they observe clear signs of normality after removing seasonal effects and trend. Following these findings, the authors propose a stochastic model for the temperature dynamics driven by a fractional Brownian motion. In particular, [Brody, Syroka and Zervos (2002)] propose the following model for temperature variations

$$T(t) = \Lambda(t) + X^H(t), \tag{6.14}$$

where $\Lambda(t)$ is as before a deterministic function modelling the trend and seasonality in temperature, and $X^H$ is a *fractional* Ornstein-Uhlenbeck process with dynamics

$$dX^H(t) = -\alpha X^H(t)\,dt + \sigma\,dB^H(t). \tag{6.15}$$

Here, $\alpha$ and $\sigma$ are two positive constants, and the process $B^H(t)$ is a fractional Brownian motion with Hurst coefficient $H \in (0,1)$. We remark that [Brody, Syroka and Zervos (2002)] consider coefficients $\alpha$ and $\sigma$ in (6.15) to be deterministic functions of time in their general model setup. Moreover, they combine the seasonality function and a mean reversion level in the Ornstein-Uhlenbeck process $X^H(t)$. However, we choose to follow the temperature models used in this book. Since [Brody, Syroka and Zervos (2002)] assume constant coefficients in their empirical analysis, we also apply this property for simplicity.

For a given $H \in (0,1)$, fractional Brownian motion $B^H(t)$ is a stochastic process with continuous sample paths starting in zero, where $B^H(t)$ is a zero-mean Gaussian random variable for all $t \geq 0$, such that

$$\mathbb{E}[B^H(t)B^H(s)] = \frac{1}{2}\left(t^{2H} + s^{2H} - |t-s|^{2H}\right). \tag{6.16}$$

In the case of $H = 0.5$ we obtain a Brownian motion, that is, $B^{0.5}(t) = B(t)$. If the Hurst coefficient $H > 0.5$, the increments of fractional Brownian motion will be positively correlated, and the paths will be smoother than those for Brownian motion. On the other hand, if $H < 0.5$, the increments are negatively correlated, and the paths will be rougher. [Brody, Syroka and Zervos (2002)] estimate a Hurst coefficient to be $H \approx 0.61$ for London data, meaning that the temperatures fluctuate smoother than Brownian motion. For $H > 0.5$, the fractional Brownian motion is long-range dependent in the sense that,

$$\sum_{n=1}^{\infty} \mathbb{E}\left[B^H(1)(B^H(n+1) - B^H(n))\right] = \infty.$$

For Brownian motion, the above sum of expectations will be zero as a consequence of its independent increment property.

Fractional Brownian motion can be represented as a Wiener integral with respect to a deterministic kernel function. It holds that

$$B^H(t) = \int_0^t g(t,s)\, dB(s), \tag{6.17}$$

where

$$g(t,s) = c_H (t-s)^{H-1/2}\, {}_2F_1\left(\frac{1}{2} - H, H - \frac{1}{2}, H + \frac{1}{2}, 1 - \frac{t}{s}\right) \tag{6.18}$$

and

$$c_H = \sqrt{\frac{2H\Gamma(\frac{3}{2} - H)}{\Gamma(H + \frac{1}{2})\Gamma(2 - 2H)}}.$$

Here, $\Gamma(\cdot)$ is the Gamma function and ${}_2F_1$ is the hypergeometric function (see Chapter 15 in [Abramowitz and Stegun (1965)]). Note that the representation of fractional Brownian motion in (6.17) links $B^H(t)$ to the BSS models considered in Sect. 6.2. In fact, fractional Brownian motion can be viewed as a particular extension of the BSS models into a non-stationary framework, where the kernel function $g$ is *not* in a stationary form $g(t-s)$. We remark that as long as $H \neq 0.5$, $B^H(t)$ is neither a semimartingale nor a Markov process.

The fractional Ornstein-Uhlenbeck process $X^H(t)$ is defined as a solution to a stochastic differential equation. A natural guess motivated by the Brownian motion case of $H = 0.5$, suggests that the solution to (6.15) is

$$X^H(t) = X^H(0)e^{-\alpha t} + \int_0^t \sigma e^{-\alpha(t-s)}\, dB^H(s). \tag{6.19}$$

But, in order for this to make sense, the meaning of integration with respect to $B^H(t)$ must be provided. Moreover, given such a definition, it has to be proven that $X^H(t)$ in (6.19) is the (unique) solution to (6.15). We next discuss these questions in more detail.

A tempting approach to define the stochastic integral with respect to fractional Brownian motion is to exploit the representation of it as a Wiener integral. However, this approach to a theory for fractional stochastic integration is not straightforward as the integrand $g(t,s)$ in (6.17) depends explicitly on $t$, and not only on $s$. The authors in [Brody, Syroka and Zervos (2002)] suggest to apply white noise analysis, making use of the Wick product in the definition (see the appendix in [Brody, Syroka and Zervos (2002)] for details). On the other hand, [Alos, Mazet and Nualart (2001)] introduce an anticipative stochastic integral for general Gaussian processes which is rather simple for deterministic integrands. We present their definition of fractional stochastic integration, which overcomes the difficulty with the kernel function $g$ depending on $t$.

Suppose that $H > 0.5$, the case of interest in temperature modelling. For a real-valued Borel measurable and bounded function $h(s)$ on $0 \le s \le t$, we define the integral operator

$$G^*(h)(t, s) = \int_s^t h(u) \frac{\partial g}{\partial u}(u, s) \, du \, . \tag{6.20}$$

By differentiation of $g$ in (6.18) we find that

$$\frac{\partial g}{\partial t}(u, s) = c_H \left( H - \frac{1}{2} \right) \left( \frac{s}{u} \right)^{\frac{1}{2} - H} (u - s)^{H - \frac{3}{2}} \, .$$

Thus, we get

$$G^*(h)(t, s) = c_H \left( H - \frac{1}{2} \right) s^{\frac{1}{2} - H} \int_s^t (u^{H - \frac{1}{2}} h(u))(u - s)^{(H - \frac{1}{2}) - 1} \, du \, . \tag{6.21}$$

Remark that in [Alos, Mazet and Nualart (2001)], the operator $G^*$ is denoted by $K_H^*$. We prefer to use $G^*$ to indicate the link to the kernel function $g$ in the fractional Brownian motion. We note that this operator may be associated with the right-sided fractional Riemann-Liouville integral of order $\nu \in (0, 1)$ of an integrable function $f$ on $[0, t]$, which is defined for almost every $s \in [0, t]$ by

$$I_{t-}^\nu f(s) = \frac{(-1)^{-\nu}}{\Gamma(\nu)} \int_s^t (u - s)^{\nu - 1} f(u) \, du \, . \tag{6.22}$$

Hence, we can express the operator $G^*$ as

$$G^*(h)(t, s) = c_H \left( H - \frac{1}{2} \right) \frac{\Gamma(H - \frac{1}{2})}{(-1)^{H - \frac{1}{2}}} I_{t-}^{H - \frac{1}{2}} h_{H - \frac{1}{2}}(s) \, ,$$

where we have introduced the notation $h_\nu(u) = u^\nu h(u)$. With the operator $G^*$ at hand, we can define integration of a deterministic function $h$ with respect to fractional Brownian motion of Hurst coefficient $H > 0.5$.

Following [Alos, Mazet and Nualart (2001)], define the integral of $h$ with respect to $B^H(t)$ by

$$X(t) = \int_0^t h(s) \, dB^H(s) = \int_0^t G^*(h)(t, s) \, dB(s) \, . \tag{6.23}$$

The integral on the right-hand side is a standard stochastic integral with respect to Brownian motion, as long as

$$\int_0^t (G^*(h)(t, s))^2 \, ds < \infty \, . \tag{6.24}$$

Since $h$ is bounded, this holds true. We observe that the fractional integral $X(t)$ in (6.23) becomes normally distributed, with mean zero and variance given by (6.24).

In the next proposition, we show the solution of the fractional Ornstein-Uhlenbeck equation.

**Proposition 6.1.** *Suppose that $H > 0.5$. Then, the stochastic process $X^H(t)$ defined in (6.19) is a solution to the fractional Ornstein-Uhlenbeck equation (6.15).*

**Proof.** Without loss of generality, we assume that $X^H(0) = 0$ and $\sigma = 1$. We calculate $X^H(t) + \alpha \int_0^t X^H(s)\,ds$, and show that it is equal $B^H(t)$. First, we find by the stochastic Fubini theorem (see [Protter (1990)]) that

$$\alpha \int_0^t X^H(s)\,ds = \alpha \int_0^t \int_0^s G^*(e^{-\alpha(s-\cdot)})(s,u)\,dB(u)\,ds$$

$$= \alpha \int_0^t \int_u^t G^*(e^{-\alpha(s-\cdot)})(s,u)\,ds\,dB(u)\,.$$

From the Fubini theorem of Lebesgue integration (see [Folland (1984)]), it holds that,

$$\alpha \int_u^t G^*(e^{-\alpha(s-\cdot)})(s,u)\,ds = \alpha \int_u^t \int_u^s e^{-\alpha(s-v)}\frac{\partial g}{\partial v}(v,u)\,dv\,ds$$

$$= \alpha \int_u^t \int_v^t e^{-\alpha(s-v)}\,ds\frac{\partial g}{\partial v}(v,u)\,dv$$

$$= -\int_u^t e^{-\alpha(t-v)}\frac{\partial g}{\partial v}(v,u)\,dv + \int_u^t \frac{\partial g}{\partial v}(v,u)\,dv$$

$$= -G^*(e^{-\alpha(t-\cdot)})(t,u) + g(t,u)\,.$$

The last equality follows from the fact that $g(u,u) = 0$ when $H > 0.5$. Thus, we find

$$X^H(t) + \alpha \int_0^t X^H(s)\,ds = \int_0^t g(t,u)\,dB(u) = B^H(t)\,,$$

and the proof is complete. $\qquad\square$

By the Gaussianity of the stochastic fractional integral, we find that $X^H(t)$ is normally distributed, with mean $X^H(0)\exp(-\alpha t)$ and variance

$$\mathrm{Var}(X^H(t)) = \int_0^t (G^*(e^{-\alpha(t-\cdot)})(t,s))^2\,ds\,. \tag{6.25}$$

This is the starting point of [Brody, Syroka and Zervos (2002)] in pricing temperature derivatives written on the CDD and HDD indices. The prices are derived at current time zero.

Let us analyze the pricing of temperature forwards in the fractional model, starting with a simple CAT futures contract. Applying the definitions, we find that

$$X^H(\tau) = X^H(0)e^{-\alpha\tau} + \sigma \int_0^\tau G^*(e^{-\alpha(\tau-\cdot)})(\tau, s)\, dB(s)$$

$$= X^H(0)e^{-\alpha\tau} + \sigma \int_0^t G^*(e^{-\alpha(\tau-\cdot)})(\tau, s)\, dB(s)$$

$$+ \sigma \int_t^\tau G^*(e^{-\alpha(\tau-\cdot)})(\tau, s)\, dB(s).$$

But,

$$G^*(e^{-\alpha(\tau-\cdot)})(\tau, s) = e^{-\alpha(\tau-t)}G^*(e^{-\alpha(t-\cdot)})(t, s) + \int_t^\tau e^{-\alpha(\tau-u)}\frac{\partial g}{\partial u}(u, s)\, du.$$

Hence,

$$X^H(\tau) = X^H(t)e^{-\alpha(\tau-t)} + \sigma \int_0^t \int_t^\tau e^{-\alpha(\tau-u)}\frac{\partial g}{\partial u}(u, s)\, du\, dB(s)$$

$$+ \sigma \int_t^\tau G^*(e^{-\alpha(\tau-\cdot)})(\tau, s)\, dB(s).$$

Next, we do a Girsanov transform (recall theory in Sect. 5.2 when specializing to the case with non-state dependent changes)

$$d\tilde{B}(t) = \frac{\theta_0(t)}{\sigma}\, dt + dB(t),$$

where $\theta_0(t)$ is a bounded measurable function. Then, the $Q^\theta$-dynamics of $X^H(t)$ is

$$X^H(\tau) = X^H(t)e^{-\alpha(\tau-t)} + e^{-\alpha(\tau-t)}\int_0^t G^*(e^{-\alpha(t-\cdot)})(t, s)\theta_0(s)\, ds$$

$$- \int_0^\tau G^*(e^{-\alpha(\tau-\cdot)})(\tau, s)\theta_0(s)\, ds$$

$$+ \sigma \int_0^t \int_t^\tau e^{-\alpha(\tau-u)}\frac{\partial g}{\partial u}(u, s)\, du\, d\tilde{B}(s)$$

$$+ \sigma \int_t^\tau G^*(e^{-\alpha(\tau-\cdot)})(\tau, s)\, d\tilde{B}(s). \tag{6.26}$$

We are now ready to derive the price dynamics of a CAT future.

**Proposition 6.2.** *Suppose $T(t)$ is given by (6.14) with $H > 0.5$. Then the CAT futures price at time $t \leq \tau_1$ with measurement period $[\tau_1, \tau_2]$ is given by*

$$F_{CAT}(t, \tau_1, \tau_2) = \int_{\tau_1}^{\tau_2} \Lambda(\tau) \, d\tau + (T(t) - \Lambda(t)) \frac{1}{\alpha} \left( e^{-\alpha(\tau_1 - t)} - e^{-\alpha(\tau_2 - t)} \right)$$

$$+ \frac{1}{\alpha} \int_t^{\tau_2} \left( G^*(e^{-\alpha(\tau_2 - \cdot)})(\tau_2, s) - g(\tau_2, s) \right) \theta_0(s) \, ds$$

$$- \frac{1}{\alpha} \int_t^{\tau_1} \left( G^*(e^{-\alpha(\tau_1 - \cdot)})(\tau_1, s) - g(\tau_1, s) \right) \theta_0(s) \, ds$$

$$+ \frac{\sigma}{\alpha} \int_0^t \left( \int_t^{\tau_1} e^{-\alpha(\tau_1 - u)} \frac{\partial g}{\partial u}(u, s) \, du - g(\tau_1, s) \right) dB(s)$$

$$- \frac{\sigma}{\alpha} \int_0^t \left( \int_t^{\tau_2} e^{-\alpha(\tau_2 - u)} \frac{\partial g}{\partial u}(u, s) \, du - g(\tau_2, s) \right) dB(s).$$

***Proof.*** From (6.26) it follows that

$$\mathbb{E}_\theta[X^H(\tau) \mid \mathcal{F}_t] = X^H(t) e^{-\alpha(\tau - t)} + e^{-\alpha(\tau - t)} \int_0^t G^*(e^{-\alpha(t - \cdot)})(t, s) \theta_0(s) \, ds$$

$$- \int_0^\tau G^*(e^{-\alpha(\tau - \cdot)})(\tau, s) \theta_0(s) \, ds$$

$$+ \sigma \int_0^t \int_t^\tau e^{-\alpha(\tau - u)} \frac{\partial g}{\partial u}(u, s) \, du \, d\widetilde{B}(s)$$

by adaptedness and the independent increment property of Brownian motion. Next, inserting $B$ from the Girsanov transform, and doing some simple algebra, we reach

$$\mathbb{E}_\theta[X^H(\tau) \mid \mathcal{F}_t] = X^H(t) e^{-\alpha(\tau - t)} - \int_t^\tau G^*(e^{-\alpha(\tau - \cdot)})(\tau, s) \theta_0(s) \, ds$$

$$+ \sigma \int_0^t \int_t^\tau e^{-\alpha(\tau - u)} \frac{\partial g}{\partial u}(u, s) \, du \, dB(s).$$

Next, by appealing to the Fubini Theorem, we find

$$\int_{\tau_1}^{\tau_2} \int_t^\tau G^*(e^{-\alpha(\tau - \cdot)})(\tau, s) \theta_0(s) \, ds \, d\tau$$

$$= \int_t^{\tau_1} \int_{\tau_1}^{\tau_2} G^*(e^{-\alpha(\tau - \cdot)})(\tau, s) \, d\tau \theta_0(s) \, ds$$

$$+ \int_{\tau_1}^{\tau_2} \int_s^{\tau_2} G^*(e^{-\alpha(\tau - \cdot)})(\tau, s) \, d\tau \theta_0(s) \, ds.$$

Moreover, it follows that,

$$\int_{\tau_1}^{\tau_2} G^*(e^{-\alpha(\tau-\cdot)}(\tau,s)\,d\tau = \frac{1}{\alpha}G^*(e^{-\alpha(\tau_1-\cdot)}(\tau_1,s) - \frac{1}{\alpha}G^*(e^{-\alpha(\tau_2-\cdot)})(\tau_2,s)$$

$$+ \frac{1}{\alpha}(g(\tau_2,s) - g(\tau_1,s)).$$

Similarly,

$$\int_{s}^{\tau_2} G^*(e^{-\alpha(\tau-\cdot)})(\tau,s)\,d\tau = -\frac{1}{\alpha}G^*(e^{-\alpha(\tau_2-\cdot)})(\tau_2,s) + \frac{1}{\alpha}g(\tau_2,s),$$

where we used the fact that $g(s,s) = 0$ as long as $H > 0.5$. Collecting terms, we find that

$$\int_{\tau_1}^{\tau_2}\int_t^\tau G^*(e^{-\alpha(\tau-\cdot)})(\tau,s)\theta_0(s)\,ds\,d\tau$$

$$= \frac{1}{\alpha}\int_t^{\tau_2}\left(G^*(e^{-\alpha(\tau_2-\cdot)})(\tau_2,s) - g(\tau_2,s)\right)\theta_0(s)\,ds$$

$$- \frac{1}{\alpha}\int_t^{\tau_1}\left(G^*(e^{-\alpha(\tau_1-\cdot)})(\tau_1,s) - g(\tau_1,s)\right)\theta_0(s)\,ds.$$

An analogous computation, using the stochastic Fubini Theorem (see [Protter (1990)]), shows that

$$\int_{\tau_1}^{\tau_2}\int_0^t\int_t^\tau e^{-\alpha(\tau-u)}\frac{\partial g}{\partial u}(u,s)\,du\,dB(s)\,d\tau$$

$$= \int_0^t\int_{\tau_1}^{\tau_2}\int_t^\tau e^{-\alpha(\tau-u)}\frac{\partial g}{\partial u}(u,s)\,du\,d\tau\,dB(s)$$

$$= \frac{1}{\alpha}\int_0^t\int_t^{\tau_1} e^{-\alpha(\tau_1)}\frac{\partial g}{\partial u}(u,s)\,du\,dB(s)$$

$$- \frac{1}{\alpha}\int_0^t\int_t^{\tau_2} e^{-\alpha(\tau_2-u)}\frac{\partial g}{\partial u}(u,s)\,du\,dB(s)$$

$$+ \frac{1}{\alpha}\int_0^t g(\tau_2,s) - g(\tau_1,s)\,dB(s).$$

Hence, the result follows.                                              □

We observe that the CAT futures price is not only depending on the current temperature and the market price of risk, but also has an additional stochastic component depending on the path of the Brownian motion up to current time $t$. This is a reflection of the memory in fractional Brownian motion.

In Prop. 6.2, we derived the futures price under the assumption that the information in the market is given by the filtration $\mathcal{F}_t$, the one generated by

Brownian motion. Admittedly, in practice, given that the temperature is following a fractional Ornstein-Uhlenbeck process, it would be more natural to condition on the filtration generated by the fractional Brownian motion $\mathcal{F}_t^H$. The fractional Brownian motion can be observed from the paths of $X^H(t)$, whereas Brownian motion must be filtered out from the paths of the fractional Brownian motion. We refer to [Benth (2003)] for a detailed analysis on dynamical pricing of temperature derivatives based on fractional Brownian motion, using white noise theory.

# Chapter 7

# Options on temperature and wind

In this chapter we price and hedge options written on weather forward contracts. We investigate European call and put options on temperature and wind forwards. Options on temperature forwards are offered for trade at the CME, while wind forward options were introduced but the US Futures Exchange was closed. A novel aspect of weather markets, namely the possibility to hedge weather risk playing on the spatial dependency of weather factors, is introduced in this Chapter. We call it geographical hedging. Geographical hedging implies using weather contracts at various locations to hedge the weather risk at a given adjacent location.

## 7.1 Options on temperature futures

As we have discussed in Sect. 1.2, European call and put options on the temperature futures are traded at the CME. In this Section we analyze the pricing and hedging of such plain vanilla options.

To this end, we denote by $C_{\mathrm{Ind}}(t, \tau, K, \tau_1, \tau_2)$ the price of a call option at time $t \geq 0$ with exercise time $\tau \geq t$ at the strike price $K$ on a temperature futures based on the index Ind measured over $[\tau_1, \tau_2]$, where Ind can take the value CAT, CDD or HDD, and $\tau \leq \tau_1$. A call option pays at exercise the difference between the underlying and the strike $K$, as long as this difference is positive, that is, the payoff function at time $\tau$ is

$$\max(F_{\mathrm{Ind}}(\tau, \tau_1, \tau_2) - K, 0).$$

A put option with strike price $K$ at exercise time $\tau$ pays

$$\max(K - F_{\mathrm{Ind}}(\tau, \tau_1, \tau_2), 0),$$

and we denote this price at time $t \leq \tau$ by $P_{\mathrm{Ind}}(t, \tau, K, \tau_1, \tau_2)$.

As in the discussion in Sect. 5.1, the no-arbitrage price of a call option is given by

$$C_{\text{Ind}}(t, \tau, K, \tau_1, \tau_2) = e^{-r(\tau-t)} \mathbb{E}_Q[\max(F_{\text{Ind}}(\tau, \tau_1, \tau_2) - K, 0) \,|\, \mathcal{F}_t], \quad (7.1)$$

where $r > 0$ is the constant risk-free interest rate and $Q \sim P$ is an equivalent martingale measure, that is, an equivalent probability such that the process $t \mapsto F_{\text{Ind}}(t, \tau_1, \tau_2)$ for $t \leq \tau$ is a $Q$-martingale. We refer to [Duffie (1992)] for a further analysis on the pricing of call options.

Similarly, we find that the no-arbitrage price of a put option is

$$P_{\text{Ind}}(t, \tau, K, \tau_1, \tau_2) = e^{-r(\tau-t)} \mathbb{E}_Q[\max(K - F_{\text{Ind}}(\tau, \tau_1, \tau_2), 0) \,|\, \mathcal{F}_t]. \quad (7.2)$$

From Sect. 5.3 we have that $F_{\text{Ind}}(t, \tau_1, \tau_2)$ is a martingale with respect to the equivalent probabilities $Q^\theta$ introduced in Sect. 5.2, and we naturally use these as the pricing measures for call and put options.

Recall the put-call parity.

**Proposition 7.1.** *It holds that*

$$C_{Ind}(t, \tau, K, \tau_1, \tau_2) - P_{Ind}(t, \tau, K, \tau_1, \tau_2)$$
$$= e^{-r(\tau-t)} F_{Ind}(t, \tau_1, \tau_2) - e^{-r(\tau-t)} K.$$

**Proof.**   Observe that

$$\max(x - K, 0) - \max(K - x, 0) = x - K$$

for all $x \in \mathbb{R}$. Inserting $F_{\text{Ind}}(\tau, \tau_1, \tau_2)$ for $x$, and taking expectations, yield

$$\mathbb{E}_\theta \left[\max(F_{\text{ind}}(\tau, \tau_1, \tau_2) - K, 0)\right] - \mathbb{E}_\theta \left[\max(K - F_{\text{ind}}(\tau, \tau_1, \tau_2), 0)\right]$$
$$= \mathbb{E}_\theta[F_{\text{Ind}}(\tau, \tau_1, \tau_2) - K \,|\, \mathcal{F}_t].$$

Using the martingale property of $F_{\text{Ind}}$ and discounting gives the result.  $\square$

By the put-call parity, we can price put options by calls. Hence, we focus our analysis on call options written on temperature futures.

Consider pricing and hedging of call options on CAT futures. The price dynamics can be derived directly using the explicit dynamics of $F_{\text{CAT}}(t, \tau_1, \tau_2)$ under $Q^\theta$.

**Proposition 7.2.** *The price of a call option at time $t \geq 0$ with exercise time $\tau \geq t$ and strike price $K$ written on a CAT futures with measurement period $[\tau_1, \tau_2]$, $\tau \leq \tau_1$, is*

$$C_{CAT}(t, \tau, K, \tau_1, \tau_2) = e^{-r(\tau-t)} \Big( (F_{CAT}(t, \tau_1, \tau_2) - K)\Phi(d(t, \tau, \tau_1, \tau_2, K))$$
$$+ \Sigma_{CAT}(t, \tau, \tau_1, \tau_2)\Phi'(d(t, \tau, \tau_1, \tau_2)) \Big),$$

*where*

$$\Sigma^2_{CAT}(t, \tau, \tau_1, \tau_2) = \int_t^\tau \mathbf{b}' C_\theta(\tau_2 - s, \tau_1 - s) \mathbf{e}_p \mathbf{e}'_p C'_\theta(\tau_2 - s, \tau_1 - s) \mathbf{b} \sigma^2(s) \, ds \,.$$

*Here, $C_\theta$ is defined in (5.20), $\Phi$ is the cumulative standard normal probability distribution and*

$$d(t, \tau, \tau_1, \tau_2, K) = \frac{F_{CAT}(t, \tau_1, \tau_2) - K}{\Sigma_{CAT}(t, \tau, \tau_1, \tau_2)} \,.$$

**Proof.** From Prop. 5.3 we have that

$$F_{\mathrm{CAT}}(\tau, \tau_1, \tau_2) = F_{\mathrm{CAT}}(t, \tau_1, \tau_2) + \int_t^\tau \mathbf{b}' C_\theta(\tau_2 - s, \tau_1 - s) \mathbf{e}_p \sigma(s) \, d\widetilde{B}(s) \,.$$

Thus, $F_{\mathrm{CAT}}(\tau, \tau_1, \tau_2)$ conditioned on $\mathcal{F}_t$ is normally distributed under $Q^\theta$, with mean $F_{\mathrm{CAT}}(t, \tau_1, \tau_2)$ and variance $\Sigma^2_{\mathrm{CAT}}(t, \tau, \tau_1, \tau_2)$ given by the Itô isometry. Hence, we find that

$$\mathbb{E}_\theta \left[ \max(F_{\mathrm{CAT}}(\tau, \tau_1, \tau_2) - K, 0) \mid \mathcal{F}_t \right]$$
$$= \mathbb{E}_\theta \left[ \max(x - K + Z \Sigma_{\mathrm{CAT}}(t, \tau, \tau_1, \tau_2), 0) \right]_{x = F_{\mathrm{CAT}}(t, \tau_1, \tau_2)} \,,$$

where $Z$ is standard normally distributed. But,

$$\mathbb{E}_\theta \left[ \max(x - K + Z \Sigma_{\mathrm{CAT}}(t, \tau, \tau_1, \tau_2), 0) \right]$$
$$= \frac{1}{\sqrt{2\pi}} \int_{\frac{K - x}{\Sigma_{\mathrm{CAT}}(t, \tau, \tau_1, \tau_2)}}^\infty (x - K + < \Sigma_{\mathrm{CAT}}(t, \tau, \tau_1, \tau_2)) \mathrm{e}^{-z^2/2} \, dz$$
$$= (x - K) \Phi \left( \frac{x - K}{\Sigma_{\mathrm{CAT}}(t, \tau, \tau_1, \tau_2)} \right)$$
$$+ \Sigma_{\mathrm{CAT}}(t, \tau, \tau_1, \tau_2) \Phi' \left( \frac{x - K}{\Sigma_{\mathrm{CAT}}(t, \tau, \tau_1, \tau_2)} \right) \,.$$

Here, $\Phi'(y)$ is the derivative of $\Phi$, the probability density function of a standard normal distribution. Hence, the result follows. $\square$

This option price is the analogue of the Black-76 formula for call options on futures (see [Black (1976)], and also Prop. 7.6). We note here that since the CAT futures dynamics is naturally on an arithmetic scale, and not a geometric Brownian motion, the pricing rule will become slightly different than the Black-76 formula. From Prop. 5.3 we find the volatility of $F_{\mathrm{CAT}}(t, \tau_1, \tau_2)$ to be

$$\sigma_{\mathrm{CAT}}(t, \tau_1, \tau_2) = \mathbf{b}' C_\theta(\tau_2 - t, \tau_1 - t) \mathbf{e}_p \sigma(t) \,.$$

But then

$$\Sigma^2_{\mathrm{CAT}}(t, \tau, \tau_1, \tau_2) = \int_t^\tau \sigma^2_{\mathrm{CAT}}(s, \tau_1, \tau_2) \, ds \,.$$

This is in analogy with the Black-76 formula, where the volatility of the futures enters into the price as the integrated variance from current time until maturity (see Prop. 9.1 in [Benth, Šaltytė Benth and Koekebakker (2008)]).

The delta-hedge ratio, also called simply the *delta* of the call option, gives the number of underlying CAT futures contracts necessary to hedge the option. The delta is given as the derivative of the call option price with respect to the CAT futures.

**Proposition 7.3.** *The delta-hedge ratio is given as*

$$\frac{\partial C_{CAT}(t, \tau, K, \tau_1, \tau_2)}{\partial F_{CAT}(t, \tau_1, \tau_2)} = e^{-r(\tau-t)} \Phi(d(t, \tau, \tau_1, \tau_2, K)),$$

*where $d(t, \tau, \tau_1, \tau_2, K)$ is defined in Prop. 7.2.*

**Proof.**    A direct differentiation gives

$$\frac{\partial C_{\text{CAT}}(t, \tau, K, \tau_1, \tau_2)}{\partial F_{\text{CAT}}(t, \tau_1, \tau_2)}$$

$$= e^{r(\tau-t)} \Big( \Phi(d(t, \tau, \tau_1, \tau_2, K))$$

$$+ (F_{\text{CAT}}(t, \tau_1, \tau_2) - K)\Phi'(d(t, \tau, \tau_1, \tau_2, K))\Sigma_{\text{CAT}}^{-1}(t, \tau, \tau_1, \tau_2)$$

$$+ \Sigma_{\text{CAT}}(t, \tau, \tau_1, \tau_2)\Phi''(d(t, \tau, \tau_1, \tau_2, K))\Sigma_{\text{CAT}}^{-1}(t, \tau, \tau_1, \tau_2) \Big)$$

$$= e^{-r(\tau-t)} \Big( \Phi(d(t, \tau, \tau_1, \tau_2, K))$$

$$+ d(t, \tau, \tau_1, \tau_2, K)\Phi'(d(t, \tau, \tau_1, \tau_2)) + \Phi''(d(t, \tau, \tau_1, \tau_2, K)) \Big).$$

Since for any $x \in \mathbb{R}$, $x\Phi'(x) + \Phi''(x) = 0$, the result follows.    □

As the cumulative distribution function gives a number between zero and one, we should hold a fraction of the underlying CAT futures contract in the hedge. In fact, if $F_{\text{CAT}}(t, \tau_1, \tau_2)$ is far bigger than $K$, such that $d(t, \tau, \tau_1, \tau_2, K)$ becomes large, $\Phi$ will be close to one. Hence, we will be close to a long position in a CAT futures. On the other hand, if $F_{\text{CAT}}(t, \tau_1, \tau_2)$ is much smaller than $K$, $d(t, \tau, \tau_1, \tau_2, K)$ will be largely negative and consequently $\Phi(d(t, \tau, \tau_1, \tau_2, K))$ close to zero. In that case, there will be almost no exposure in the CAT futures. Note that the discounting term is reducing the fraction yielded by $\Phi(d(t, \tau, \tau_1, \tau_2, K))$, however, getting closer to one as $t$ approaches the exercise time $\tau$.

For call options on CDD futures we are not able to derive an explicit "Black-76-like" formula for the price. However, the price may be expressed suitably as an expectation of a functional of a standard normally distributed random variable, feasible for Monte Carlo simulations.

**Proposition 7.4.** *The price of a call option at time $t \geq 0$ with exercise time $\tau \geq t$ and strike price $K$ written on a CDD futures with measurement period $[\tau_1, \tau_2]$, $\tau \leq \tau_1$, is*

$$C_{CDD}(t, \tau, K, \tau_1, \tau_2)$$

$$= e^{-r(\tau - t)} \mathbb{E} \left[ \max \left( \int_{\tau_1}^{\tau_2} \Sigma_\theta(\tau, s) \right. \right.$$

$$\left. \left. \times \Psi \left( \frac{m_\theta(t, s, \mathbf{x}) - c + Z\Sigma(t, \tau, s)}{\Sigma_\theta(\tau, s)} \right) ds - K, 0 \right) \right]_{\mathbf{x} = \mathbf{X}(t)},$$

*where $Z$ is a standard normally distributed random variable and*

$$\Sigma^2(t, \tau, s) = \int_t^\tau \mathbf{b}' \exp(A_\theta(s - u)) \mathbf{e}_p \mathbf{e}_p' \exp(A_\theta'(s - u)) \mathbf{b} \sigma^2(u) \, du .$$

**Proof.** From Prop. 5.4 we have

$$F_{CDD}(\tau, \tau_1, \tau_2) = \int_{\tau_1}^{\tau_2} \Sigma_\theta(\tau, s) \Psi \left( \frac{m_\theta(\tau, s, \mathbf{X}(\tau)) - c}{\Sigma_\theta(\tau, s)} \right) ds .$$

Notice from Lemma 5.1 that for $\tau \geq t$

$$\mathbf{X}(\tau) = \exp(A_\theta(\tau - t)) \mathbf{X}(t) + \int_t^\tau \exp(A_\theta(\tau - u)) \mathbf{e}_p \theta_0(u) \, du$$

$$+ \int_t^\tau \exp(A_\theta(\tau - u)) \mathbf{e}_p \sigma(u) \, d\widetilde{B}(u) .$$

From the definition of $m_\theta(\tau, s, \mathbf{x})$ in (5.17) it holds that

$$m_\theta(\tau, s, \mathbf{X}(\tau)) = \Lambda(s) + \mathbf{b}' \exp(A_\theta(s - \tau)) \exp(A_\theta(\tau - t)) \mathbf{X}(t)$$

$$+ \mathbf{b}' \exp(A_\theta(s - \tau)) \int_t^\tau \exp(A_\theta(\tau - u)) \mathbf{e}_p \sigma(u) \, d\widetilde{B}(u)$$

$$+ \mathbf{b}' \exp(A_\theta(s - \tau)) \int_t^\tau \exp(A_\theta(\tau - u)) \mathbf{e}_p \theta_0(u) \, du$$

$$+ \int_\tau^s \mathbf{b}' \exp(A_\theta(s - u)) \mathbf{e}_p \theta_0(u) \, du$$

$$= \Lambda(s) + \mathbf{b}' \exp(A_\theta(s - t)) \mathbf{X}(t)$$

$$+ \int_t^s \mathbf{b}' \exp(A_\theta(s - t)) \mathbf{e}_p \theta_0(u) \, du$$

$$+ \int_t^\tau \mathbf{b}' \exp(A_\theta(s - u)) \mathbf{e}_p \sigma(u) \, d\widetilde{B}(u)$$

$$= m_\theta(t, s, \mathbf{X}(t)) + \int_t^\tau \mathbf{b}' \exp(A_\theta(s - u)) \mathbf{e}_p \sigma(u) \, d\widetilde{B}(u) .$$

By the independent increment property of Brownian motion, we have that the stochastic integral above is independent of $\mathcal{F}_t$ with respect to the probability $Q^\theta$. Furthermore, $\mathbf{X}(t)$ is $\mathcal{F}_t$-adapted. Hence, by the Markov property, we find

$$
\begin{aligned}
C_{\text{CDD}}&(t, \tau, K, \tau_1, \tau_2) \\
&= e^{-r(\tau-t)} \mathbb{E}_\theta \left[ \max \left( F_{\text{CDD}}(\tau, \tau_1, \tau_2) - K, 0 \right) \mid \mathcal{F}_t \right] \\
&= e^{-r(\tau-t)} \mathbb{E} \left[ \max \left( \int_{\tau_1}^{\tau_2} \Sigma_\theta(\tau, s) \right. \right. \\
&\qquad\qquad \left. \left. \times \Psi \left( \frac{m_\theta(t, s, \mathbf{x}) - c + Z\Sigma(t, \tau, s)}{\Sigma_\theta(\tau, s)} \right) ds - K, 0 \right) \right]_{\mathbf{x}=\mathbf{X}(t)}.
\end{aligned}
$$

Thus, the Proposition follows. □

We observe that $\Sigma_\theta(t, s) = \Sigma(t, s, s)$ in the Proposition above. To valuate the call option on the CDD futures, we simply simulate a number of independent outcomes $z_1, \ldots, z_M$ of the standard normally distributed variable $Z$, and compute the empirical mean

$$
\begin{aligned}
e^{r(\tau-t)} & C_{\text{CDD}}(t, \tau, K, \tau_1, \tau_2) \\
&\approx \frac{1}{M} \sum_{m=1}^{M} \left( \int_{\tau_1}^{\tau_2} \Sigma_\theta(\tau, s) \Psi \left( \frac{m_\theta(t, s, \mathbf{x}) - c + z_m \Sigma(t, \tau, s)}{\Sigma_\theta(\tau, s)} \right) ds - K, 0 \right)^+,
\end{aligned}
$$

where we insert $\mathbf{x} = \mathbf{X}(t)$, and denote $(x)^+ = \max(x, 0)$. This is a straightforward Monte Carlo simulation of the price. The integral with respect to time must be evaluated using Riemann approximation. We recall from the definition of the CDD futures (see Subsect. 1.2.1) that our time integral is a mathematical approximation of a discrete sum of daily sampled CDD indices. Hence, a daily sampled Riemann integral is the correct numerical approximation to the integral. It turns out that we need a very high number of samples $M$ to ensure a good approximation as it converges as $1/\sqrt{M}$ in standard deviation. There exists a number of methods to speed up Monte Carlo simulations using variance reducing techniques or quasi-Monte Carlo methods. We refer the interested reader to [Glasserman (2003)] for an extensive introduction. Note that the numerical integral also has to be re-evaluated for each sample $z_m$, which increases the computational burden of this approach significantly.

The call option price dynamics on CDD futures does *not* depend explicitly on the current CDD futures price $F_{\text{CDD}}(t, \tau_1, \tau_2)$. The option price will be a non-linear function of $\mathbf{b}' \exp(A_\theta(s - t))\mathbf{X}(t)$, meaning that the price depends on the current states $\mathbf{X}(t)$ of the temperature. Let us analyze this dependency a bit closer.

From Prop. 5.4 we find

$$\frac{\partial}{\partial s}F_{\mathrm{CDD}}(t, \tau_1, s) = \Sigma_\theta(t, s)\Psi\left(\frac{m_\theta(t, s, \mathbf{X}(t)) - c}{\Sigma_\theta(t, s)}\right),$$

for $s \in (\tau_1, \tau_2]$. Since $\Psi'(x) = \Phi(x) > 0$, $\Psi(x)$ is a monotonely increasing function and thus has an inverse. Hence,

$$\mathbf{b}'\exp(A_\theta(s - t))\mathbf{X}(t) = c - \Lambda(s) - \int_t^s \mathbf{b}'\exp(A_\theta(s - u))\mathbf{e}_p\theta_0(u)\,du$$

$$+ \Sigma_\theta(t, s)\Psi^{-1}\left(\frac{\frac{\partial}{\partial s}F_{\mathrm{CDD}}(t, \tau_1, s)}{\Sigma_\theta(t, s)}\right).$$

From this we conclude that the call option price on a CDD futures contract with measurement period $[\tau_1, \tau_2]$ depends on *all* CDD futures with measurement periods $[\tau_1, s]$, for $s \in (\tau_1, \tau_2]$. In fact, the call price depends on the sensitivity of these CDD futures prices with respect to the terminal time of the measurement period. However, the conclusion remains that $C_{\mathrm{CDD}}(t, \tau, K, \tau_1, \tau_2)$ is *not* a function of $F_{\mathrm{CDD}}(t, \tau_1, \tau_2)$ only.

Let us proceed with investigating the hedging strategy of a call option on a CDD futures. As we have just argued, the option price is not a function of the CDD futures price, and thus the hedging strategy cannot be computed by the delta. We must resort to the martingale representation formula, or more specifically the Clark-Ocone formula. As we have that

$$C_{\mathrm{CDD}}(\tau, \tau, K, \tau_1, \tau_2) = \max(F_{\mathrm{CDD}}(\tau, \tau_1, \tau_2) - K, 0), \qquad (7.3)$$

the Clark-Ocone formula (see [Karatzas, Ocone and Li (1991)]) says that there exists an Itô integrable process $t \mapsto \xi(t, \tau), t \le \tau$, such that

$$C_{\mathrm{CDD}}(\tau, \tau, K, \tau_1, \tau_2) = \mathbb{E}_\theta\left[C_{\mathrm{CDD}}(\tau, \tau, K, \tau_1, \tau_2)\right] + \int_0^\tau \xi(t, \tau)\,d\widetilde{B}(t), \quad (7.4)$$

with

$$\xi(t, \tau) = \mathbb{E}_\theta\left[D_t C_{\mathrm{CDD}}(\tau, \tau, K, \tau_1, \tau_2) \mid \mathcal{F}_t\right]. \qquad (7.5)$$

The definition of the process $\xi(t, \tau)$ involves the *Malliavin derivative*, $D_t$, of a random variable. For more on the theory of the Malliavin derivative, the interested reader is recommended to consult the monograph [Malliavin and Thalmaier (2006)]. From Prop. 5.5, we recall that

$$dF_{\mathrm{CDD}}(t, \tau_1, \tau_2) = \sigma(t)\psi(t, \tau_1, \tau_2, \mathbf{X}(t))\,d\widetilde{B}(t),$$

with

$$\psi(t, \tau_1, \tau_2, \mathbf{X}(t)) = \int_{\tau_1}^{\tau_2} (\mathbf{b}' \exp(A_\theta(s-t))\mathbf{e}_p) \, \Phi\left(\frac{m_\theta(t, s, \mathbf{X}(t)) - c}{\Sigma_\theta(t, s)}\right) ds \,.$$

(7.6)

Hence, substituting $d\widetilde{B}(t)$ with $dF_{\mathrm{CDD}}(t, \tau_1, \tau_2)$ in (7.4), we find

$$C_{\mathrm{CDD}}(\tau, \tau, K, \tau_1, \tau_2) = \mathbb{E}_\theta \left[C_{\mathrm{CDD}}(\tau, \tau, K, \tau_1, \tau_2)\right]$$

$$+ \int_0^\tau \frac{\xi(t, \tau)}{\sigma(t)\psi(t, \tau_1, \tau_2, \mathbf{X}(t))} \, dF_{\mathrm{CDD}}(t, \tau_1, \tau_2) \,.$$

Then the integrand in the stochastic integral above gives the hedging position in the CDD futures contract. To compute this position, we need to derive $\xi(t, \tau)$. The next Lemma states an explicit form for the Malliavin derivative of the payoff function.

**Lemma 7.1.** *The Malliavin derivative of $C_{CDD}(\tau, \tau, K, \tau_1, \tau_2)$ is*

$$D_t C_{CDD}(\tau, \tau, K, \tau_1, \tau_2) = \mathbf{1}\left(F_{CDD}(\tau, \tau_1, \tau_2) > K\right)$$

$$\times \sigma(t) \int_{\tau_1}^{\tau_2} \Phi\left(\frac{m_\theta(\tau, s, \mathbf{X}(\tau)) - c}{\Sigma_\theta(\tau, s)}\right) \mathbf{b}' \exp(A_\theta(s-t))\mathbf{e}_p \, ds \,,$$

*where $\mathbf{1}(A)$ is the indicator function on the set $A$.*

**Proof.** From the chain rule of the Malliavin derivative, we have

$$D_t C_{\mathrm{CDD}}(\tau, \tau, K, \tau_1, \tau_2) = \mathbf{1}\left(F_{\mathrm{CDD}}(\tau, \tau_1, \tau_2) > K\right) D_t F_{\mathrm{CDD}}(\tau, \tau_1, \tau_2) \,.$$

Following Prop. 5.4 and repeatedly applying the chain rule, we find

$$D_t F_{\mathrm{CDD}}(\tau, \tau_1, \tau_2) = \int_{\tau_1}^{\tau_2} \Sigma_\theta(\tau, s) \Psi'\left(\frac{m_\theta(\tau, s, \mathbf{X}(\tau)) - c}{\Sigma_\theta(\tau, s)}\right)$$

$$\times \frac{1}{\Sigma_\theta(\tau, s)} D_t m_\theta(\tau, s\mathbf{X}(\tau)) \, ds \,.$$

Recall that $\Psi(x) = x\Phi(x) + \phi(x)$ and therefore $\Psi'(x) = \Phi(x)$, where $\Phi(x)$ is the cumulative distribution function of a standard normal random variable with density $\phi(x)$.

From the definition of $m_\theta(\tau, s, \mathbf{x})$ in (5.17) it holds that

$$D_t m_\theta(\tau, s, \mathbf{X}(\tau)) = D_t \left(\Lambda(s) + \mathbf{b}'e^{A_\theta(s-\tau)}\mathbf{X}(\tau) + \int_\tau^s \mathbf{b}'e^{A_\theta(s-u)}\mathbf{e}_p\theta_0(u) \, du\right)$$

$$= D_t \mathbf{b}'e^{A_\theta(s-\tau)}\mathbf{X}(\tau) \,,$$

since the Malliavin derivative of deterministic functions is zero. From Lemma 5.1 we therefore find that

$$D_t \mathbf{b}'e^{A_\theta(s-\tau)}\mathbf{X}(\tau) = D_t \int_0^\tau (\mathbf{b}'e^{A_\theta(s-u)}\mathbf{e}_p)\sigma(u) \, d\widetilde{B}(u)$$

$$= (\mathbf{b}'e^{A_\theta(s-t)}\mathbf{e}_p)\sigma(t) \,.$$

Hence, the proof is complete. $\qquad\square$

Note that the payoff functional in (7.3) is not a composition of a smooth function and a random variable, and therefore, strictly speaking, we cannot use the chain rule for the Malliavin derivative in the proof above. However, we can extend the proof by applying a limiting argument using smooth functions which approximates the max-operator. We refer to the monograph [Di Nunno, Øksendal and Proske (2009)] for details and similar applications of Malliavin Calculus to finance.

From the above Lemma, we have

$$\xi(t,\tau) = \mathbb{E}_\theta \left[ D_t C_{\mathrm{CDD}}(\tau,\tau,K,\tau_1,\tau_2) \,|\, \mathcal{F}_t \right]$$

$$= \mathbb{E}_\theta \left[ \mathbf{1}\left(F_{\mathrm{CDD}}(\tau,\tau_1,\tau_2) > K\right) \sigma(t) \int_{\tau_1}^{\tau_2} \mathbf{b}' e^{A_\theta (s-t)} \mathbf{e}_p \right.$$

$$\left. \times \Phi\left( \frac{m_\theta(\tau,s,\mathbf{X}(\tau)) - c}{\Sigma_\theta(\tau,s)} \right) ds \,\Big|\, \mathcal{F}_t \right]$$

$$= \sigma(t) \int_{\tau_1}^{\tau_2} \left( \mathbf{b}' e^{A_\theta (s-t)} \mathbf{e}_p \right) \mathbb{E}_\theta \left[ \mathbf{1}\left(F_{\mathrm{CDD}}(\tau,\tau_1,\tau_2) > K\right) \right.$$

$$\left. \times \Phi\left( \frac{m_\theta(\tau,s,\mathbf{X}(\tau)) - c}{\Sigma_\theta(\tau,s)} \right) \,\Big|\, \mathcal{F}_t \right] ds .$$

From Lemma 5.1, $\mathbf{X}(\tau)$ can be expressed in terms of $\mathbf{X}(t)$ and an Itô integral independent of $\mathcal{F}_t$. Hence, following similar arguments as for the call option price above, we can express the conditional expectation for $\xi(t,\tau)$ as a function of $\mathbf{X}(t)$. Therefore, there exists a function $\widetilde{\xi}(t,\tau,\tau_1,\tau_2,\mathbf{X}(t))$ such that

$$\xi(t,\tau) = \sigma(t)\widetilde{\xi}(t,\tau,\tau_1,\tau_2,\mathbf{X}(t)) .$$

We conclude that

$$C_{\mathrm{CDD}}(\tau,\tau,K,\tau_1,\tau_2) = \mathbb{E}_\theta \left[ C_{\mathrm{CDD}}(\tau,\tau,K,\tau_1,\tau_2) \right]$$

$$+ \int_0^\tau \frac{\widetilde{\xi}(t,\tau,\tau_1,\tau_2,\mathbf{X}(t))}{\psi(t,\tau_1,\tau_2,\mathbf{X}(t))} \, dF_{\mathrm{CDD}}(t,\tau_1,\tau_2) .$$

The "delta hedge" is then expressible as the ratio

$$\frac{\widetilde{\xi}(t,\tau,\tau_1,\tau_2,\mathbf{X}(t))}{\psi(t,\tau_1,\tau_2,\mathbf{X}(t))} .$$

Unfortunately, this ratio cannot be written analytically, but must be computed by an expectation dependent on $\mathbf{X}(t)$. Monte Carlo simulations can also be applied for this purpose. Note that the hedge ratio depends explicitly on the vector $\mathbf{X}(t)$, and not only on the current temperature and CDD futures price.

## 7.2 Options on wind speed futures

The pricing of options on wind speed futures is analogous to the CDD temperature futures case in the previous section. We consider here a call option written on a wind speed futures settled on the Nordix wind speed with measurement period $[\tau_1, \tau_2]$. Then the following holds for the price of the call.

**Proposition 7.5.** *The price at time $t \geq 0$ of a call option with exercise time $\tau \geq t$ and strike $K$ on a wind speed futures with measurement period $[\tau_1, \tau_2]$, $\tau \leq \tau_1$, is*

$$C_N(t, \tau, K, \tau_1, \tau_2)$$
$$= e^{-r(\tau-t)} \mathbb{E} \Big[ \max\Big( \int_{\tau_1}^{\tau_2} \Xi\left(\tau, s, m_\theta(t, s, \mathbf{x}) + Z\Sigma(t, \tau, s)\right) ds$$
$$+ c(\tau_1, \tau_2) - K, 0 \Big) \Big]_{\mathbf{x}=\mathbf{X}(t)},$$

*where*

$$\Xi(\tau, s, x) = M_{\frac{1}{\lambda}} \left(1 + \lambda x, \lambda^2 \Sigma_\theta(\tau, s)\right)$$

*for $\lambda \in (0, 1]$ and*

$$\Xi(\tau, s, x) = \exp\left(x + \frac{1}{2}\Sigma_\theta^2(\tau, s)\right)$$

*for $\lambda = 0$. Here, $M_{1/\lambda}(a, b^2)$ is the $(1/\lambda)$th moment of a normally distributed random variable with mean $a$ and variance $b^2$. Furthermore, $\Sigma(t, \tau, s)$ is defined in Prop. 7.4 and $Z$ is a standard normally distributed random variable.*

**Proof.** We have from (5.32) that

$$F_N(\tau, \tau_1, \tau_2) = c(\tau_1, \tau_2) + \int_{\tau_1}^{\tau_2} f_N(\tau, s) \, ds$$

with $f_N(\tau, s)$ defined in Prop. 5.8. Note that, from Prop. 5.8, $f_N(\tau, s)$ for $s \geq \tau$ can be written as

$$f_N(\tau, s) = \Xi(\tau, s, m_\theta(\tau, s, \mathbf{X}(\tau))).$$

Using the same sequence of calculations as in the proof of Prop. 7.4, we have

$$m_\theta(\tau, s, \mathbf{X}(\tau)) = m_\theta(t, s, \mathbf{X}(t)) + \int_t^\tau \mathbf{b}' \exp(A_\theta(s - u))\mathbf{e}_p \sigma(u) \, d\widetilde{B}(u).$$

Applying the Markov property after using that Brownian motion has independent increments and $\mathbf{X}(t)$ is $\mathcal{F}_t$-adapted, yields

$$C_N(t, \tau, K, \tau_1, \tau_2)$$
$$= e^{-r(\tau-t)} \mathbb{E}_\theta \left[ \max \left( F_N(\tau, \tau_1, \tau_2) - K, 0 \right) \mid \mathcal{F}_t \right]$$
$$= e^{-r(\tau-t)} \mathbb{E}_\theta \left[ \max \left( \int_{\tau_1}^{\tau_2} f_N(\tau, s) \, ds + c(\tau_1, \tau_2) - K, 0 \right) \mid \mathcal{F}_t \right]$$
$$= e^{-r(\tau-t)} \mathbb{E} \left[ \max \left( \int_{\tau_1}^{\tau_2} \Xi \left( \tau, s, m_\theta(t, s, \mathbf{x}) + Z \Sigma(t, \tau, s) \right) ds \right. \right.$$
$$\left. \left. + c(\tau_1, \tau_2) - K, 0 \right) \right]_{\mathbf{x} = \mathbf{X}(t)}.$$

The result follows. ☐

Similar to call options on CDD futures, the price $C_N(t, \tau, K, \tau_1, \tau_2)$ is not a function of the current wind speed futures price $F_N(t, \tau_2, \tau_2)$. But it is dependent on the states $\mathbf{X}(t)$ in the CARMA process defining the current wind speed.

To calculate the price $C_N(t, \tau, K, \tau_1, \tau_2)$, we may use Monte Carlo simulations as discussed for options on CDD futures above. A tempting alternative to Monte Carlo pricing frequently used in computational finance is a transform-based valuation method (see [Carr and Madan (1998)]). Such a method is based on the Fourier transform of the payoff function and the inverse Fourier transform. In the present situation, we can compute the Fourier transform of the function

$$z \mapsto \max \left( \int_{\tau_1}^{\tau_2} \Xi \left( \tau, s, m_\theta(t, s, \mathbf{x}) + z \Sigma(t, \tau, s) \right) ds + c(\tau_1, \tau_2) - K, 0 \right),$$

and apply a transform-based valuation. This is not a straightforward task, as the function is defined via an integral over time to be computed numerically. This in turn will require derivation of the Fourier transform by a numerical integration. Finally, notice that the function is not integrable, hence it is necessary to apply the Fourier transform of some dampened version of it. In conclusion, it is highly questionable if a transform-based valuation of the call option will be efficient in the case of options on wind speed futures. We come back to the Fourier method for pricing derivatives when we analyze options on precipitation in Chapter 8.

Let us discuss a "quick fix" to efficient pricing of call options in the case of $\lambda = 0$. By Prop. 5.8 we find

$$\frac{df_N(t, s)}{f_N(t, s)} = \mathbf{b}' \exp(A_\theta(s - t)) \mathbf{e}_p \sigma(t) \, d\widetilde{B}(t)$$

for $s \geq t$. The volatility of the geometric Brownian motion dynamics of the process $t \mapsto f_N(t, s)$ is $s$-dependent, thus it is easy to show that the stochastic process

$$t \mapsto \int_{\tau_1}^{\tau_2} f_N(t, s) \, ds$$

does not become a geometric Brownian motion. Therefore $F_N(t, \tau_1, \tau_2)$ is *not* a geometric Brownian motion. However, it is possible to introduce an approximate geometric Brownian motion dynamics of $F_N(t, \tau_1, \tau_2)$ by simply defining a volatility $\Sigma_N(t, \tau_1, \tau_2)$ as

$$\Sigma_N^2(t, \tau_1, \tau_2) = \int_{\tau_1}^{\tau_2} \mathbf{b}' \exp(A_\theta(s - t)) \mathbf{e}_p \mathbf{e}_p' \exp(A_\theta'(s - t)) \mathbf{b} \, ds \sigma^2(t) \,, \quad (7.7)$$

and assuming that the dynamics of $F_N(t, \tau_1, \tau_2)$ is

$$\frac{dF_N(t, \tau_1, \tau_2)}{F_N(t, \tau_1, \tau_2)} = \Sigma_N(t, \tau_1, \tau_2) \, d\widetilde{B}(t) \,. \quad (7.8)$$

The validity of such approximations is analyzed theoretically and empirically in [Benth (2010)] in the case of Ornstein-Uhlenbeck processes, that is, in the case of CAR processes with $p = 1$. We claim that also for higher-order CARMA processes this is a good approximation, leading to reasonable prices of the call.

Given the approximative dynamics in (7.8), we can easily obtain a Black-76 formula for the price.

**Proposition 7.6.** *The price at time $t \geq 0$ of a call option with exercise time $\tau \geq t$ and strike price $K$ written on a wind forward with dynamics as in (7.8) is*

$$C_N(t, \tau, K, \tau_1, \tau_2) = \mathrm{e}^{-r(\tau - t)} F_N(t, \tau_1, \tau_2) \Phi(d_1) - \mathrm{e}^{-r(\tau - t)} K \Phi(d_2) \,,$$

*where*

$$d_1 = \frac{\ln F_N(t, \tau_1, \tau_2) - \ln K + \frac{1}{2} \int_t^\tau \Sigma_N^2(s, \tau_1, \tau_2) \, ds}{\sqrt{\int_t^\tau \Sigma_N^2(s, \tau_1, \tau_2) \, ds}}$$

*and*

$$d_2 = \frac{\ln F_N(t, \tau_1, \tau_2) - \ln K - \frac{1}{2} \int_t^\tau \Sigma_N^2(s, \tau_1, \tau_2) \, ds}{\sqrt{\int_t^\tau \Sigma_N^2(s, \tau_1, \tau_2) \, ds}} \,.$$

**Proof.** By the geometric Brownian motion dynamics in (7.8), we find from Itô's Formula, Thm. 4.1, that

$$F_N(\tau, \tau_1, \tau_2) = F_N(t, \tau_1, \tau_2) \exp\left(-\frac{1}{2}\int_t^\tau \Sigma_N^2(s, \tau_1, \tau_2)\, ds \right.$$
$$\left. + \int_t^\tau \Sigma_N(s, \tau_1, \tau_2)\, d\widetilde{B}(s)\right).$$

Since $F_N(t, \tau_1, \tau_2)$ is $\mathcal{F}_t$-adapted, and the stochastic integral is independent of $\mathcal{F}_t$ due to the independent increment property of Brownian motion, we obtain that

$$C_N(t, \tau, K, \tau_1, \tau_2)$$
$$= e^{-r(\tau-t)}\mathbb{E}_\theta\left[\max\left(F_N(\tau, \tau_1, \tau_2) - K\right) \mid \mathcal{F}_t\right]$$
$$= e^{-r(\tau-t)}\mathbb{E}\left[\max\left(x\exp\left(-\frac{1}{2}\widetilde{\Sigma}^2 + Z\widetilde{\Sigma}\right) - K, 0\right)\right]_{x=F_N(t,\tau_1,\tau_2)},$$

where $Z$ is a standard normally distributed random variable and $\widetilde{\Sigma}$

$$\widetilde{\Sigma}^2 = \int_t^\tau \Sigma_N^2(s, \tau_1, \tau_2)\, ds.$$

Since

$$\mathbb{E}\left[\max\left(x\exp\left(-\frac{1}{2}\widetilde{\Sigma}^2 + Z\widetilde{\Sigma}\right) - K, 0\right)\right]$$
$$= \frac{1}{\sqrt{2\pi}}\int_{(\ln(x/K)+\frac{1}{2}\widetilde{\Sigma}^2)/\widetilde{\Sigma}}^\infty \left(xe^{-\frac{1}{2}\widetilde{\Sigma}^2+z\widetilde{\Sigma}} - K\right)e^{-z^2/2}\, dz,$$

the result follows by a straightforward integration using the properties of the normal probability density function. $\square$

We observe that this price dynamics for the call option on wind forwards allows for a simple computation of the delta-hedge ratio, and thus it is easy to find an approximative hedging strategy in the underlying wind speed futures for the option. It holds that

$$\frac{\partial C_N(t, \tau, K, \tau_1, \tau_2)}{\partial F_N(t, \tau_1, \tau_2)} = e^{-r(\tau-t)}\Phi(d_1)$$

$$+ e^{-r(\tau-t)}\Phi'(d_1)\frac{1}{\sqrt{\int_t^\tau \Sigma_N(s, \tau_1, \tau_2)\,ds}}\frac{1}{F_N(t, \tau_1, \tau_2)}$$

$$- e^{-r(\tau_t)}K\Phi'(d_2)\frac{1}{\sqrt{\int_t^\tau \Sigma_N(s, \tau_1, \tau_2)\,ds}}\frac{1}{F_N(t, \tau_1, \tau_2)}$$

$$= e^{-r(\tau-t)}\Phi(d_1) + e^{-r(\tau-t)}\frac{1}{\sqrt{\int_t^\tau \Sigma_N(s, \tau_1, \tau_2)\,ds}}$$

$$\times \left(\Phi'(d_1) - \frac{K}{F_N(t, \tau_1, \tau_2)}\Phi'(d_2)\right)$$

$$= e^{-r(\tau-t)}\Phi(d_1)\,.$$

This is, not unexpectedly, similar to the hedge ratio for calls on CAT futures.

Motivated by the so-called Heath-Jarrow-Morton approach to forward rate modelling in fixed-income markets (see [Heath, Jarrow and Morton (1992)]), one can view the dynamics in (7.8) as a possibility to model the wind speed futures price dynamics directly. In other words, $F_N(t, \tau_1, \tau_2)$ is not derived from a wind speed model, but modelled directly as a stochastic process. Based on wind speed futures price data, the volatility $\Sigma_N(t, \tau_1, \tau_2)$ can be estimated and in this way the price dynamics is fully specified. The problem with this approach is that it may be hard to recover the wind speed dynamics from it, and therefore it can be impossible to link changes in wind speed to the forward prices.

Finally, we remark that we can try the same approximative approach on the dynamics of wind speed futures when $\lambda \in (0, 1]$. The method is also relevant in the case of CDD temperature futures. We come back to this in Sect. 9.1, where such geometric Brownian motion approximations to pricing derivatives using a so-called utility indifference methods are applied.

## 7.3   Geographical hedging

Weather is a spatially varying phenomena, where different geographical locations experience different states of weather. Locations close to each other may have very similar weather, whereas distant locations may differ significantly in temperature, precipitation or wind conditions. Of course, geographical differences as rivers, mountains or coast lines affect and determine weather systems. In this section we focus on how the dependency in weather conditions between different locations may be exploited financially to hedge weather exposure.

As a simple example, we consider a summer holiday resort. Obviously, such a resort is dependent on good weather, in the sense of little rain and high temperatures. The resort runs a financial risk if temperatures are low (for example below 20°C), or due to rainfall over a longer period (for example, several consecutive days with rain). To hedge this exposure to weather risk, the resort may enter into call or put options, or forward contracts, on rainfall and/or temperature. In Europe, say, there are many holiday resorts on the Mediterranean coastline, however, not many weather derivatives are written for this area. The problem is therefore that even though the holiday resort may know their weather risk exposure, there are no contracts devoted to offset this risk in the weather markets.

The holiday resort may of course contact a financial intermediary that may sell an appropriate derivative, but this may be very expensive compared to the traded products in a market, the CME temperature futures and options, say. Alternatively, the resort may create a basket of *market traded* weather derivatives that partly hedge their risk. In this way, the holiday resort will exploit the dependency in weather conditions between their own location, and locations where derivatives are traded. For example, a resort on the French Riviera may use Paris, Madrid and Rome temperature futures and options, to create a hedge, assuming that there is a significant degree of correlation between the resort, and the three cities. In order to analyze such hedges based on derivatives in different locations, spatial models for weather are required. We refer to hedging strategies exploiting the spatial dependency structure of weather as *geographical hedges*.

In the above example, we focused on how weather exposure in *one* specific location could be hedged using a basket of derivatives in other locations. This problem can be extended to hedging the exposure to weather in a whole geographical area. For example, many power utilities have their main market in a restricted geographic area. The demand for the electricity production is dependent on temperature in that area. A low average temperature in the area of interest may lead to high demand for heating, and thus high income for the company, while higher temperatures reduce the profit. To hedge this temperature risk, the utility could enter into a basket of temperature futures and options for different locations, inside and outside of the area, which makes up a hedge against the average temperature risk. In this case the average temperature is hedged in a domain, by using contracts traded in the temperature in a finite number of discrete locations.

A final example before we enter into more quantitative considerations, may be to revisit the financial intermediary that wants to offer tailormade derivatives to specific locations. This intermediary may be an insurance company or a bank, willing to specify weather insurance in terms of derivatives at locations not present at CME. Of course, this intermediary will not sit with naked positions, taking on all the risk in the derivatives sold. In order to price the contracts correctly, making sure that the own risk is covered, the intermediary will try to bundle together a hedge using traded contracts at an exchange. This again leads to geographical hedging, since a basket of weather derivatives that will offset as much as possible of the exposure in the contract of the intermediary at some specific location is constructed. In this sense, we use a geographical strategy to hedge an option or some exotic weather derivative.

In the rest of this section we focus on geographical hedging of temperature risk. First, we present a simple toy model for the spatial-temporal evolution of temperature, motivated from the temperature modelling in Chapter 3. Next, we apply this to study the problem of hedging temperature risk using market-traded futures. The analysis is based on [Barth, Benth and Potthoff (2011)].

### 7.3.1   *A simple spatial-temporal model for temperature*

Recall from the analyses in Chapter 3 that the temperature dynamics in most stations in Lithuania are well described by an AR(3) process with seasonal mean and seasonal volatility. This observation motivated the introduction of seasonal CARMA models applied in Chapter 4. A simple trend-surface model for the parameters of the CARMA model and the seasonality function described the spatial structure of the temporal parameters satisfactory. Furthermore, the spatial dependency structure of the residuals at different locations was well modelled by a spherical or exponential correlation function.

These considerations motivate the following temperature dynamics in continuous time and space. Let $T(t, \mathbf{s})$ be the temperature at time $t \geq 0$ and location $\mathbf{s} = (x, y) \in \mathbb{R}^2$. Assume that

$$T(t, \mathbf{s}) = \Lambda(t, \mathbf{s}) + \mathbf{b}'(\mathbf{s})\mathbf{X}(t, \mathbf{s}),\qquad(7.9)$$

where at each location **s** this is a seasonal CARMA($p, q$) model, meaning that all parameters in the CARMA model of temperature are location-dependent. More specifically, we let

$$\mathbf{X}(s, \mathbf{s}) = e^{A(\mathbf{s})(s-t)}\mathbf{X}(t, \mathbf{s}) + \int_t^s e^{A(\mathbf{s})(s-u)}\mathbf{e}_p\sigma(u, \mathbf{s})\,dB(u, \mathbf{s})\,. \qquad (7.10)$$

Here, for each **s**, $A(\mathbf{s})$ is a matrix for the CAR process as defined in (4.2). Hence, we assume that the coefficients $\alpha_1, \ldots, \alpha_p$ in the last row are location-dependent. Moreover, we assume that the deterministic seasonal functions $\Lambda$ and $\sigma$ are location-dependent as well. All the dependencies on **s** in $\Lambda$, $\sigma$ and $A$ are assumed to be deterministic and smooth.

We have introduced the location-dependent Brownian motion $B(t, \mathbf{s})$. By this we understand a Brownian motion process $t \mapsto B(t, \mathbf{s})$ for each fixed location **s**, and a Gaussian random field for each fixed time $t$, that is $\mathbf{s} \mapsto B(t, \mathbf{s})$ is a Gaussian random field. In fact, this yields that $(t, \mathbf{s}) \mapsto B(t, \mathbf{s})$ is a mean-zero Gaussian random field in time and space, with variance-covariance function

$$\text{Cov}\left(B(t, \mathbf{s}), B(\widetilde{t}, \widetilde{\mathbf{s}})\right) = \min(t, \widetilde{t})\rho(\mathbf{s}, \widetilde{\mathbf{s}})\,, \qquad (7.11)$$

for some non-negative function $\rho$ on $\mathbb{R}^2$, $\mathbf{s}, \widetilde{\mathbf{s}} \in \mathbb{R}^2$. As $B(t, \mathbf{s})$ is a Brownian motion for each spatial location, its variance is then given by $t$. Hence, $\rho(\mathbf{s}, \mathbf{s}) = 1$ which implies that $\rho$ is in fact a *correlation* function. Notice that the variance-covariance structure completely defines the Gaussian random field.

From the analysis in Chapter 3, a stationary correlation function was suggested as an appropriate choice in the case of Lithuanian temperatures, which means that

$$\rho(\mathbf{s}, \widetilde{\mathbf{s}}) = \rho(|\mathbf{s} - \widetilde{\mathbf{s}}|)\,. \qquad (7.12)$$

In fact, empirical analysis demonstrated anisotropy in the spatial dependency structure, meaning that the correlation depends differently on the distance in $x$-direction than in the $y$-direction. Exponential or spherical anisotropic correlation functions as defined in (3.14) and (3.15) were proposed as the optimal choices when analyzing Lithuanian data.

A mathematically rigorous way to define the spatial-temporal process $B(t, \mathbf{s})$ is to interpret $t \mapsto B(t, \cdot)$ as a Hilbert space valued Brownian motion, that is, a process which takes values in some Hilbert space $H$. There is a well-developed theory for Hilbert space valued stochastic processes, and we refer to the classical reference [Da Prato and Zabczyk (1992)] on the subject. In [Barth, Benth and Potthoff (2011)] this approach is worked out in detail for the current setting.

### 7.3.2   *Computation of the optimal geographical hedge*

We suppose that the market offers liquid trade in temperature futures at $n$ locations, $\mathbf{s}_1, \ldots, \mathbf{s}_n$. The futures in these locations are settled against temperature indices $\text{Ind}(\tau_1^i, \tau_2^i, \mathbf{s}_i)$, $i = 1, \ldots, n$, where the indices may be of CAT, HDD or CDD type. We may even have a mixture of such indices, for example, some CATs and some CDDs. The measurement periods $[\tau_1^i, \tau_2^i]$ may also be different for each location. We have included $\mathbf{s}_i$ in the notation for the index at location $i$ to highlight the dependency on spatial location.

Consider now an investor who seeks temperature risk protection in a location $\mathbf{s}$ which does not coincide with any of the locations $\mathbf{s}_i, i = 1, \ldots, n$. We focus on the case where an investor wishes to swap a floating temperature index $\text{Ind}(\tau_1, \tau_2, \mathbf{s})$ against a fixed, that is, wants to enter a temperature futures contract at location $\mathbf{s}$ on the index Ind with measurement period $[\tau_1, \tau_2]$. As there are no such futures traded, the investor's goal is to find a best representative in the market by entering a *portfolio* of traded futures. To reach this goal, the investor will take advantage of the spatial dependency in creating the portfolio.

As an example, let us assume that the investor is located so that there are two neighbouring cities in a reasonable distance from $\mathbf{s}$ where the market trades temperature futures. This may be the city of Hamburg in Germany, where CME is trading temperature futures in Essen and Berlin. One may imagine a strong correlation between the temperature in Hamburg and Berlin and Essen. The investor could create a *synthetic* futures contract for Hamburg by putting together a portfolio of temperature futures in Essen and Berlin. If a CAT futures for June in Hamburg is of interest, the investor could focus on the two CAT futures measured over June in Essen and Berlin and find a trading strategy in those which give the "optimal" synthetic CAT futures in Hamburg.

The problem is of course what is meant by the "optimal" synthetic futures. As the correlation between the temperatures in Hamburg and Berlin and Essen are not perfect, it is only possible to obtain an approximative synthetic futures. Here, we define the "optimal" synthetic futures contract as the one minimizing the variance of the error.

To present it in mathematical terms, let $\mathbf{a}(t) \in \mathbb{R}^n$ be a vector describing the number of futures at time $t$ entered at $n$ different locations where such contracts are traded. We assume that the $n$-dimensional stochastic process $\mathbf{a}(t)$ is adapted, that is, measurable with respect to $\mathcal{F}_t$ at each time $t \geq 0$, where $\mathcal{F}_t$ is the filtration generated by $(s, \mathbf{s}) \mapsto B(s, \mathbf{s})$ for $s \leq t$ and

$\mathbf{s} \in \mathbb{R}^2$. Adaptedness of $\mathbf{a}$ is a natural condition ensuring that the investor takes actions based on the current market information. The investor seeks to find the optimal investment strategy $\widehat{\mathbf{a}}(t)$ minimizing the variance of the error defined by

$$R(t, \mathbf{a}) = \mathbb{E}\left[\left(\mathrm{Ind}(\tau_1, \tau_2, \mathbf{s}) - \sum_{i=1}^{n} \mathbf{a}_i(t)\mathrm{Ind}(\tau_1^i, \tau_2^i, \mathbf{s}_i)\right)^2 \Big| \mathcal{F}_t\right]. \quad (7.13)$$

Mathematically, we define $\widehat{\mathbf{a}}$ to be

$$\widehat{\mathbf{a}}(t) = \arg\min_{\mathbf{a}(t)} R(t, \mathbf{a}), \quad (7.14)$$

where the minimum is taken over all adapted strategies such that $R$ is finite. By entering futures positions based on the indices $\mathrm{Ind}(\tau_1^i, \tau_2^i, \mathbf{s}_i)$ according to the vector $\widehat{\mathbf{a}}(t)$, the investor obtains a synthetic futures contract written on $\mathrm{Ind}(\tau_1, \tau_2, \mathbf{s})$. It will be the futures position that best replicates the desired one.

Rewriting the functional $R(t, \mathbf{a})$, we obtain

$$R(t, \mathbf{a}) = \mathbb{E}\left[\mathrm{Ind}^2(\tau_1, \tau_2, \mathbf{s}) \,|\, \mathcal{F}_t\right]$$
$$- 2\sum_{i=1}^{n} \mathbf{a}_i(t)\mathbb{E}\left[\mathrm{Ind}(\tau_1, \tau_2, \mathbf{s})\mathrm{Ind}(\tau_1^i, \tau_2^i, \mathbf{s}_i) \,|\, \mathcal{F}_t\right]$$
$$+ \sum_{i,j=1}^{n} \mathbf{a}_i(t)\mathbf{a}_j(t)\mathbb{E}\left[\mathrm{Ind}(\tau_1^i, \tau_2^i, \mathbf{s}_i)\mathrm{Ind}(\tau_1^j, \tau_2^j, \mathbf{s}_j) \,|\, \mathcal{F}_t\right].$$

Let $C(t)$ be the $(n \times n)$-matrix with elements defined as

$$c_{ij}(t) = \frac{\mathbb{E}\left[\mathrm{Ind}(\tau_1^i, \tau_2^i, \mathbf{s}_i)\mathrm{Ind}(\tau_1^j, \tau_2^j, \mathbf{s}_j) \,|\, \mathcal{F}_t\right]}{\mathbb{E}\left[\mathrm{Ind}^2(\tau_1^i, \tau_2^i, \mathbf{s}_i) \,|\, \mathcal{F}_t\right]}, \quad (7.15)$$

for $i, j = 1, \ldots, n$. Note that the diagonal elements are all equal to one, and the matrix $C(t)$ is symmetric. Further, we introduce the vector $\mathbf{c}(t) \in \mathbb{R}^n$ with coordinates

$$c_i(t) = \frac{\mathbb{E}\left[\mathrm{Ind}(\tau_1, \tau_2, \mathbf{s})\mathrm{Ind}(\tau_1^j, \tau_2^j, \mathbf{s}_j) \,|\, \mathcal{F}_t\right]}{\mathbb{E}\left[\mathrm{Ind}^2(\tau_1^i, \tau_2^i, \mathbf{s}_i) \,|\, \mathcal{F}_t\right]}, \quad (7.16)$$

for $i = 1, \ldots, n$. Then we have the following.

**Proposition 7.7.** *The minimizer $\widehat{\mathbf{a}}(t)$ is given as the solution of the linear system*

$$C(t)\mathbf{a}(t) = \mathbf{c}(t).$$

**Proof.** The first-order condition for a minimum $\partial R/\partial a_j(t) = 0$, $j = 1, \ldots, n$, yields

$$\sum_{i=1}^{n} \mathbf{a}_i(t)\mathbb{E}\left[\mathrm{Ind}(\tau_1^i, \tau_2^i, \mathbf{s}_i)\mathrm{Ind}(\tau_1^j, \tau_2^j, \mathbf{s}_j) \mid \mathcal{F}_t\right]$$

$$= \mathbb{E}\left[\mathrm{Ind}(\tau_1, \tau_2, \mathbf{s})\mathrm{Ind}(\tau_1^j, \tau_2^j, \mathbf{s}_j) \mid \mathcal{F}_t\right].$$

Hence, the result follows. $\qquad\qquad\qquad\qquad\qquad\qquad\qquad\square$

Note that if $\mathbf{s} = \mathbf{s}_i$ and $\tau_1 = \tau_1^i$, $\tau_2 = \tau_2^i$ for some $i = 1, \ldots, n$, then the solution is $\hat{\mathbf{a}}(t) = \mathbf{e}_i$. Hence, as expected, the synthetic futures coincides with the actual futures at location $i$, $i = 1, \ldots, n$.

From the temperature model in time and space, we can compute the "covariances" $c_{ij}$ and $\mathbf{c}_i$. Next, after finding the inverse of $C(t)$ we obtain the optimal synthetic hedging portfolio. We specialize to the case of CAT futures to illustrate these steps (remarking that the CDD and HDD futures cases are not so explicit).

For two different locations $\mathbf{s}_i$ and $\mathbf{s}_j$, we find in the case of CAT futures that

$$\mathbb{E}\left[\int_{\tau_1^i}^{\tau_2^i} T(u, \mathbf{s}_i)\, du \int_{\tau_1^j}^{\tau_2^j} T(v, \mathbf{s}_j)\, dv \,\middle|\, \mathcal{F}_t\right]$$

$$= \int_{\tau_1^i}^{\tau_2^i}\int_{\tau_1^j}^{\tau_2^j} \mathbb{E}\left[T(u, \mathbf{s}_i)T(v, \mathbf{s}_j) \mid \mathcal{F}_t\right] du\, dv$$

$$= \int_{\tau_1^i}^{\tau_2^i}\int_{\tau_1^j}^{\tau_2^j} \Lambda(u, \mathbf{s}_i)\Lambda(v, \mathbf{s}_j)\, du\, dv$$

$$+ \int_{\tau_1^i}^{\tau_2^i}\int_{\tau_1^j}^{\tau_2^j} \Lambda(v, \mathbf{s}_j)\mathbf{b}'(\mathbf{s}_i)\mathbb{E}\left[\mathbf{X}(u, \mathbf{s}_i) \mid \mathcal{F}_t\right] du\, dv$$

$$+ \int_{\tau_1^i}^{\tau_2^i}\int_{\tau_1^j}^{\tau_2^j} \Lambda(u, \mathbf{s}_i)\mathbf{b}'(\mathbf{s}_j)\mathbb{E}\left[\mathbf{X}(v, \mathbf{s}_j) \mid \mathcal{F}_t\right] du\, dv$$

$$+ \mathbf{b}'(\mathbf{s}_i)\int_{\tau_1^i}^{\tau_2^i}\int_{\tau_1^j}^{\tau_2^j} \mathbb{E}\left[\mathbf{X}(u, \mathbf{s}_i)\mathbf{X}'(v, \mathbf{s}_j) \mid \mathcal{F}_t\right] du\, dv\, \mathbf{b}(\mathbf{s}_j)$$

after applying (7.9). By using (7.10), we find from $\mathcal{F}_t$-measurability and independent increment property of Brownian motion that

$$\mathbb{E}\left[\mathbf{X}(u, \mathbf{s}_i) \mid \mathcal{F}_t\right] = e^{A(\mathbf{s}_i)(u-t)}\mathbf{X}(t, \mathbf{s}_i)$$

$$+ \mathbb{E}\left[\int_t^u e^{A(\mathbf{s}_i)(u-z)}\mathbf{e}_p\sigma(z, \mathbf{s}_i)\, dB(z, \mathbf{s}_i)\right]$$

$$= e^{A(\mathbf{s}_i)(u-t)}\mathbf{X}(t, \mathbf{s}_i).$$

By using the same arguments, it follows that

$$\mathbb{E}\left[\mathbf{X}(u, \mathbf{s}_i)\mathbf{X}'(v, \mathbf{s}_j) \mid \mathcal{F}_t\right]$$
$$= e^{A(\mathbf{s}_i)(u-t)}\mathbf{X}(t, \mathbf{s}_i)\mathbf{X}'(t, \mathbf{s}_j)e^{A'(\mathbf{s}_j)(v-t)}$$
$$+ \mathbb{E}\left[\left(\int_t^u e^{A(\mathbf{s}_i)(u-z)}\mathbf{e}_p\sigma(z, \mathbf{s}_i)\,dB(z, \mathbf{s}_i)\right)\right.$$
$$\left.\left(\int_t^v e^{A(\mathbf{s}_j)(v-z)}\mathbf{e}\sigma(z, \mathbf{s}_j)\,dB(z, \mathbf{s}_j)\right)' \,\middle|\, \mathcal{F}_t\right]$$
$$= e^{A(\mathbf{s}_i)(u-t)}\mathbf{X}(t, \mathbf{s}_i)\mathbf{X}'(t, \mathbf{s}_j)e^{A'(\mathbf{s}_j)(v-t)}$$
$$+ \rho(\mathbf{s}_i, \mathbf{s}_j)\int_t^{\min(u,v)} e^{A(\mathbf{s}_i)(u-z)}\mathbf{e}_p\mathbf{e}_p'e^{A(\mathbf{s}_j)(v-z)}\sigma(z, \mathbf{s}_i)\sigma(z, \mathbf{s}_j)\,dz\,.$$

In the last step we used the correlation of the Brownian motion between two locations $\mathbf{s}_i$ and $\mathbf{s}_j$ together with the Itô isometry. In conclusion, for the case of CAT futures, we have all the ingredients required in order to compute the elements of the matrix $C(t)$ and the vector $\mathbf{c}(t)$.

For given seasonality function $\Lambda(t, \mathbf{s})$ and volatility $\sigma(t, \mathbf{s})$, an explicit computation implies an integration, that in practical situations requires the use of numerical integration techniques. As the above expression depends explicitly on the spatial correlation structure $\rho(\mathbf{s}_i, \mathbf{s}_j)$, the locations with the tradeable futures and the location of the synthetic futures will play a crucial role for how good the synthetic futures will be. We can thus compute the residual hedging error, which depends on the spatial correlation structure. We refer to [Barth, Benth and Potthoff (2011)] for numerical considerations. Summarizing, the quality of the synthetic futures will depend on the number of tradeable futures available, the distance between locations $\mathbf{s}_i$ and $\mathbf{s}$, and also the geometry of the locations. In [Barth, Benth and Potthoff (2011)], examples where the optimal synthetic futures contracts consist of both long and short positions in the various traded ones can be found.

# Chapter 8

# Modelling and pricing derivatives on precipitation

The modelling and analysis of precipitation derivatives are based on different stochastic models than the CARMA processes used for wind speed and temperature. The modelling of precipitation involves the application of independent increment processes. Independent increment processes are an extension of the class of Lévy processes, to which Brownian motion belongs. We construct a continuous-time analogue of the time series model proposed in Sect. 3.3, and apply it for pricing various derivatives contracts. The market price of risk will be introduced via the Esscher transform, which preserves the independent increment property of the driving processes. Furthermore, we derive semi-analytical pricing formulas for derivatives like swaps and options based on Fourier transform methods and properties of independent increment processes. This chapter provides the basic background required on independent increment processes. For those not familiar with these processes and wishing to learn more, we refer to [Benth, Šaltytė Benth and Koekebakker (2008)] and the references therein for an introduction in the context of energy markets. See also [Cont and Tankov (2004)] for the theory on the special case of Lévy processes and their applications to finance, and [Jacod and Shiryaev (1987)] for a complete account on independent increment processes.

## 8.1 A continuous-time model for precipitation

Before defining our stochastic model for precipitation, we give a brief introduction to the independent increment processes driving the dynamics. Our presentation is based on the introduction in [Benth, Šaltytė Benth and Koekebakker (2008)].

### 8.1.1   *A class of independent increment processes*

Let $L(t)$ be a square integrable independent increment process defined on the filtered probability space $(\Omega, \mathcal{F}, \{\mathcal{F}\}_{t\geq 0}, P)$, that is, a process with finite variance where the increments are independent, starting at zero and are stochastically continuous. We choose the version of $L$ with right-continuous paths with left-limits (RCLL). If $L(t)$ has stationary increments, then it is called a Lévy process. Stationarity of increments means that the changes in value of $L$ over a time interval, $L(t) - L(s)$ for $t > s \geq 0$ say, have a distribution only dependent on $t - s$, and not on $t$ and $s$ separately. This property rules out the possibility to have a frequency of the occurrence of rain- or snowfall dependent on the season. Independent increment processes, on the other hand, have such a flexibility. We restrict our attention to pure jump processes $L$ which are increasing. In the context of precipitation modelling, the increment of $L$ will measure the *amount* of rain or snow over some time interval. Thus, it is natural to assume $L$ to be an increasing process stating the accumulated amount of precipitation.

The *cumulant* function of an independent increment process $L$ is

$$\psi(s, t; x) = \ln \mathbb{E}\left[\exp\left(ix(L(t) - L(s))\right)\right]$$

$$= ix(\widetilde{\gamma}(t) - \widetilde{\gamma}(s)) - \frac{1}{2}x^2(C(t) - C(s))$$

$$+ \int_s^t \int_{\mathbb{R}} \left(e^{ixz} - 1 - ixz\mathbf{1}_{|z|<1}\right) \ell(dz, du)$$

for $x \in \mathbb{R}$, $t > s \geq 0$ and $\mathbf{1}_A$ being the indicator function on the (Borel) subset $A$ of $\mathbb{R}$. Here, $\widetilde{\gamma}$ is a continuous function with $\widetilde{\gamma}(0) = 0$, usually referred to as the drift, while $C(t)$ is a non-decreasing continuous function, where $C(0) = 0$ is the covariance of the continuous martingale part of $L$. As we restrict our attention to pure jump processes $L$, $C(t) = 0$ for all $t \geq 0$, and hence this term disappears from the cumulant function. We call $\ell(dz, du)$ the *compensator measure* of $L$, and it is defined as the measure

$$\ell((0, t] \times A) = \mathbb{E}\left[N((0, t] \times A)\right], \tag{8.1}$$

for $t \geq 0$ and $A$ a Borel subset in $\mathbb{R}\backslash\{0\}$. Here, $N$ is the *random jump measure* defined as

$$N((0, t] \times A) = \sum_{s \leq t} \mathbf{1}_{\Delta L(s) \in A}. \tag{8.2}$$

In the above we used the notation $\Delta L(s) = L(s) - L(s-)$ and $L(s-) = \lim_{u\uparrow s} L(u)$ which exists since $L$ is an RCLL process. Notice that $\Delta L(s)$ is the jump size at time $s$, which potentially can be zero if no jump occurs. The compensator measure satisfies

$$\ell(\{0\} \times A) = 0 \text{ and } \ell(\{t\} \times \mathbb{R}) = 0$$

for all $t \geq 0$ and $A$ a Borel subset of $\mathbb{R}\backslash\{0\}$. Moreover, for all $t \geq 0$,

$$\int_0^t \int_{\mathbb{R}} \min(1, z^2)\, \ell(dz, du) < \infty. \tag{8.3}$$

If $L$ is a Lévy process, then $\ell(dz, du) = \widetilde{\ell}(dz)\, du$, where $\widetilde{\ell}(dz)$ is the *Lévy measure*. In addition, the drift is linear, that is, of the form $\widetilde{\gamma}(t) = \widetilde{\gamma}t$. Clearly, the cumulant of a Lévy process can be expressed as $\psi(s, t, x) = \widetilde{\psi}(x)(t - s)$ for a function $\widetilde{\psi}(x)$, the cumulant of $L(1)$.

Recall that we are interested only in increasing independent increment processes, implying that the jumps can only be positive. This is reflected in a compensator measure only supported on $\mathbb{R}_+$. Moreover, an increasing function has paths of finite variation, and therefore the compensator measure integrates $z$ over the interval $[0, 1)$, and not only $z^2$ as in (8.3). Therefore, we can write

$$\int_s^t \int_{\mathbb{R}_+} \left(e^{ixz} - 1 - ixz\mathbf{1}_{|z|<1}\right) \ell(dz, du) = -ix \int_s^t \int_0^1 z\, \ell(dz, du)$$

$$+ \int_s^t \int_{\mathbb{R}_+} (e^{ixz} - 1)\, \ell(dz, du).$$

In conclusion, the cumulant function for the considered class of processes can be defined as

$$\psi(s, t, x) = ix(\gamma(t) - \gamma(s)) + \int_s^t \int_{\mathbb{R}_+} (e^{ixz} - 1)\, \ell(dz, du), \tag{8.4}$$

for a *non-decreasing* continuous function $\gamma(t)$ with $\gamma(0) = 0$. In most models of interest, the drift is zero, so from now on we also assume that $L$ is driftless, that is, $\gamma(t) = 0$ for all $t \geq 0$.

As we assume $L$ to have finite variance, we can express both the expectation and variance of $L$ via the cumulant function. The expectation of $L(t)$ is

$$\mathbb{E}\left[L(t)\right] = -i\frac{d}{dx}\psi(0, t, x)|_{x=0} = \int_0^t \int_{\mathbb{R}_+} z\, \ell(dz, du), \tag{8.5}$$

while the second moment of the increment $L(t) - L(s)$ for $t > s \geq 0$ is

$$\mathbb{E}\left[(L(t) - L(s))^2\right] = -\frac{d^2}{dx^2}\psi(s,t,x)|_{x=0} - \left(\frac{d}{dx}\psi(s,t,x)|_{x=0}\right)^2$$

$$= \int_s^t \int_{\mathbb{R}_+} z^2\,\ell(dz,du) - \left(\mathrm{i}\int_s^t \int_{\mathbb{R}_+} z\,\ell(dz,du)\right)^2.$$

$$(8.6)$$

Then the variance of $L(t) - L(s)$ becomes

$$\mathrm{Var}(L(t) - L(s)) = \int_s^t z^2\,\ell(dz,du)\,. \qquad (8.7)$$

Frequently, the expectation and variance of $L(t) - L(s)$ are directly expressed using the cumulants as

$$\mathbb{E}\left[L(t) - L(s)\right] = -\mathrm{i}\psi'(s,t,0) \qquad (8.8)$$

and

$$\mathrm{Var}(L(t) - L(s)) = -\psi''(s,t,0)\,, \qquad (8.9)$$

respectively, where $\psi^{(n)}(s,t,x) := \frac{d^n}{dx^n}\psi(s,t,x)$. Note that $\psi'(s,t,0) = \psi'(0,t,0) - \psi'(0,s,0)$.

### 8.1.2   *A stochastic model of precipitation*

Let $g : \mathbb{R}_+ \mapsto \mathbb{R}_+$ be a measurable function and consider the stochastic process

$$R(t) = \int_0^t g(s)\,dL(s)\,. \qquad (8.10)$$

The function $g$ is scaling the jumps of $L$. From now on, $R(t)$ will be the model for the cumulative amount of precipitation at time $t$ in a given location. The accumulation starts at time $t = 0$, which naturally could be chosen as January 1 in a specific year. Let us analyze the process $R(t)$.

First, we need to understand what the stochastic integral in (8.10) means. The intepretation of this is in the standard Wiener sense, as we now briefly explain.

Assume that $h$ is a simple function on $[0,t]$, that is, a function which can be expressed as

$$h(s) = \sum_{i=0}^{n-1} a_i \mathbf{1}_{[t_i,t_{i+1})}(s)\,,$$

for (positive) constants $a_i$, $i = 0, \ldots, n-1$, and a partition $\{t_i\}_{i=0,\ldots,n}$ of the interval $[0, t]$ with $t_0 = 0$ and $t_n = t$. Then the stochastic integral in (8.10) can be *defined* as

$$\int_0^t h(s)\, dL(s) = \sum_{i=0}^{n-1} a_i \Delta L(t_i),\tag{8.11}$$

where $\Delta L(t_i) = L(t_{i+1}) - L(t_i)$. By the independence of the increments of $L$ we find that the variance of this integral is

$$\mathrm{Var}\left(\int_0^t h(s)\, dL(s)\right) = \sum_{i=0}^{n-1} a_i \mathrm{Var}(\Delta L(t_i))$$

$$= \sum_{i=0}^{n-1} a_i \int_{t_i}^{t_{i+1}} \int_{\mathbb{R}_+} z^2\, \ell(dz, du).$$

This definition of the stochastic integral for simple functions suggests the following procedure for defining the integral of a general function $g$. Assume first that $g$ satisfies the integrability hypothesis

$$\int_0^T \int_{\mathbb{R}_+} g^2(u) z^2\, \ell(dz, du) < \infty,\tag{8.12}$$

for some $T < \infty$. If $g$ is bounded, then this condition holds true. Letting $t \leq T$, we sample $g$ at the partitioning times $t_i$ to introduce the simple function approximation

$$g_n(s) = \sum_{i=0}^{n-1} g(t_i) \mathbf{1}_{[t_i, t_{i+1})}(s),\tag{8.13}$$

for each $n \in \mathbb{N}$. We next define the sequence of random variables $G_n(t)$ by

$$G_n(t) = \sum_{i=0}^{n-1} g(t_i) \Delta L(t_i).\tag{8.14}$$

By assuming that the partitions $\{t_i\}_{i=0,\ldots,n}$ are nested within $n$, we have that

$$\lim_{n\to\infty} \sum_{i=0}^{n-1} \int_{t_i}^{t_{i+1}} \int_{\mathbb{R}_+} g^2(t_i) z^2\, \ell(dz, du) = \int_0^t \int_{\mathbb{R}_+} g^2(u) z^2\, \ell(dz, du).$$

Hence, it follows that $G_n(t)$ is a Cauchy sequence in the Hilbert space $L^2(\Omega, \mathcal{F}, P)$. Thus, there exists a unique limiting element in $L^2(\Omega, \mathcal{F}, P)$ called the stochastic integral of $g$ with respect to $L$

$$\int_0^t g(s)\, dL(s) = \lim_{n\to\infty} G_n(t),\tag{8.15}$$

where the limit is in $L^2(\Omega, \mathcal{F}, P)$. Moreover, it holds that

$$\text{Var}\left(\int_0^t g(s)\, dL(s)\right) = \int_0^t \int_{\mathbb{R}_+} g^2(s) z^2\, \ell(dz, ds)\,. \qquad (8.16)$$

Finally, we observe that the stochastic integral becomes a $\mathcal{F}_t$-adapted process on $[0, T]$ by construction.

We have the following integration-by-parts formula.

**Proposition 8.1.** *Assume that $g$ is continuously differentiable. Then*

$$\int_0^t g(s)\, dL(s) = g(t) L(t) - \int_0^t L(s) g'(s)\, ds\,.$$

**Proof.**   It holds that

$$\sum_{i=0}^{n-1} g(t_i)(L(t_{i+1}) - L(t_i)) = \sum_{i=0}^{n-1} (g(t_{i+1}) L(t_{i+1}) - g(t_i) L(t_i))$$

$$- \sum_{i=0}^{n-1} L(t_{i+1})(g(t_{i+1}) - g(t_i))$$

$$= g(t) L(t) - \sum_{i=0}^{n-1} L(t_{i+1})(g(t_{i+1}) - g(t_i))\,.$$

The last sum converges to the Lebesgue integral $\int_0^t L(s) g'(s)\, ds$ as $n \to \infty$, since $g'(s) L(s)$ is Lebesgue integrable on $[0, t]$ by right-continuity of $L$. This proves the proposition. $\qquad \square$

The above integration-by-parts formula can be used as a *definition* of the stochastic integral. We remark that this integration-by-parts formula may be extended to the case of more general functions $g$.

We next relate the continuous-time precipitation model $R(t)$ defined in (8.10) to the time series model proposed in Sect. 3.3. From the above definition of the stochastic integral, an Euler approximation of $R(t + \delta) - R(t)$ over a small interval $[t, t + \delta]$, $\delta > 0$, will take the form

$$R(t + \delta) - R(t) = \int_t^{t+\delta} g(s)\, dL(s) \approx g(t) \Delta_\delta L(t)\,, \qquad (8.17)$$

where $\Delta_\delta L(t) = L(t+\delta) - L(t)$. In this approximation, we may view $\Delta_\delta L(t)$ as the stochastic amount of precipitation over the short time interval from $t$ to $t + \delta$. The function $g$ is introduced in order to have a possible seasonal scaling of the amount of rain or snow.

To be more specific, consider a compound Poisson process $L(t)$ with time-dependent intensity $\lambda(t)$. Such a process can be represented as

$$L(t) = \sum_{i=1}^{N(t)} J_i, \tag{8.18}$$

where $N(t)$ is a Poisson process with intensity $\lambda(t)$ and $J_i$ are independent and identically distributed random variables. Note that the intensity of the increment $N(t) - N(s)$ is equal to $\int_s^t \lambda(u)\,du$. In order for the independent increment process to be increasing, we assume that the random variables $J_i$, measuring the jump sizes of $L$, are positive. Hence, the distribution function of $J_i$, $F_J$, is supported on $\mathbb{R}_+$. The cumulant and compensator measure for this process are stated in the next Lemma.

**Lemma 8.1.** *The compound Poisson process $L$ with time-dependent intensity $\lambda(t) > 0$ and jumps $J$ has cumulant function*

$$\psi(s, t, x) = \int_s^t \lambda(u)\,du \left( e^{\psi_J(x)} - 1 \right)$$

*for $0 \leq s < t$ and $x \in \mathbb{R}$, where $\psi_j(x)$ is the cumulant of $J$. The compensator measure is*

$$\ell(dz, du) = \lambda(u) F_J(dz)\,du,$$

*where $F_J$ is the distribution of $J$.*

**Proof.** Note that

$$L(t) - L(s) = \sum_{i=N(s)+1}^{N(t)} J_i,$$

which in distribution is equal to

$$L(t) - L(s) = \sum_{i=1}^{N(t)-N(s)} J_i.$$

Conditioning on $N(t) - N(s) = k$, we find after applying the property of the independence of $J_i$'s

$$\ln \mathbb{E}\left[ \exp(ix \sum_{i=1}^k J_i) \right] = \sum_{i=1}^k \ln \mathbb{E}\left[ \exp(ix J_i) \right] = k \psi_J(x).$$

Then

$$\ln \mathbb{E}\left[\exp(ix(L(t) - L(s)))\right] = -\int_s^t \lambda(u)\,du + \ln \sum_{k=0}^{\infty} \frac{\left(\int_s^t \lambda(u)\,du\right)^k}{k!} e^{k\psi_J(x)}$$

$$= \int_s^t \lambda(u)\,du \left(e^{\psi_J(x)} - 1\right),$$

which is the cumulant function. As

$$\mathbb{E}\left[e^{ixJ} - 1\right] = \int_{\mathbb{R}} (e^{ixz} - 1)\,F_J(dz),$$

we find that

$$\psi(s, t, x) = \int_s^t \int_{\mathbb{R}} (e^{ixz} - 1)\,F_J(dz)\lambda(u)\,du,$$

which proves the Lemma.     □

From this Lemma, we see that the compound Poisson process $L$ has no drift $\gamma$ in its cumulant. This is a typical situation.

A suggestive notation for $dL(t)$ in the case of a compound Poisson process is

$$dL(t) = J\,dN(t),$$

interpreted as follows. If $N$ has a jump over the infinitesimal time increment $[t, t + dt)$, then the process $L$ jumps with the jump size $J$. On a discrete scale, we thus can write

$$\Delta_\delta L(t) \approx J\Delta_\delta N(t). \tag{8.19}$$

Note that $N$ can have more than one jump over any time interval, but the probability of it is negligible as $\delta$ gets small.

We can interpret the model of $R(t)$ on a discrete time scale in the following way. At random times we observe the occurrence of rain or snowfall, with the amount $J$. The random times of occurrence of precipitation arrive at an intensity rate $\lambda(t)$. The time dependency allows for modelling a seasonality in the frequency of precipitation. Finally, the function $g$ scales the jumps, that is, the amount of precipitation, according to season to control for seasonality in the *amount* as well. This is in accordance with the model proposed in Sect. 3.3. Recall that we found explicit representations of $g$ and $\lambda$ for rainfall observed in Lithuania. Moreover, the jumps $J$ were modelled as exponentially distributed random variables.

Next, the cumulant function for $R(t)$ is computed.

**Proposition 8.2.** *The cumulant of $R(t)$ in* (8.10) *is given by*

$$\psi_{R(t)}(x) = \psi(0, t, xg(\cdot)),$$

*where $\psi(s, t, x)$ is defined in* (8.4) *and we interpret $\psi(0, t, xg(\cdot))$ as*

$$\psi(0, t, xg(\cdot)) = \int_0^t \int_{\mathbb{R}_+} (e^{ixg(u)z} - 1)\, \ell(dz, du).$$

*Proof.* We have, by using the definition of the stochastic integral and the Fubini-Tonelli theorem (see [Folland (1984)]) that

$$\psi_{R(t)}(x) = \ln \mathbb{E}\left[\exp(ix \int_0^t g(s)\, dL(s))\right]$$

$$= \lim_{n \to \infty} \ln \mathbb{E}\left[\prod_{i=0}^{n-1} \exp(ixg(t_i)\Delta L(t_i))\right]$$

$$= \lim_{n \to \infty} \sum_{i=0}^{n-1} \ln \mathbb{E}\left[\exp(ixg(t_i)\Delta L(t_i))\right].$$

The property of independent increments is applied in the last step. From the cumulant of $L(t) - L(s)$ in (8.4) we have

$$\psi_{R(t)}(x) = \lim_{n \to \infty} \sum_{i=0}^{n-1} \psi(t_i, t_{i+1}, xg(t_i))$$

$$= \lim_{n \to \infty} \sum_{i=0}^{n-1} \int_{t_i}^{t_{i+1}} \int_{\mathbb{R}_+} (e^{ixg(t_i)z} - 1)\, \ell(dz, du)$$

$$= \int_s^t \int_{\mathbb{R}_+} (e^{ixg(u)z} - 1)\, \ell(dz, du),$$

which proves the result. $\qquad\square$

We observe from the model that $R(t) - R(s)$ is independent of $R(s)$ for all $t > s$ due to the independent increment property of $L$. This means that the cumulant above can be used straightforwardly to find the cumulant for $R(t) - R(s)$. It holds that

$$\psi_{R(t)}(x) = \psi_{R(t)-R(s)}(x) + \psi_{R(s)}(x),$$

and hence

$$\psi_{R(t)-R(s)}(x) = \int_s^t \int_{\mathbb{R}_+} (e^{ixg(u)z} - 1)\, \ell(dz, du),$$

for the increment of $R$.

Let us go back to the example with a compound Poisson specification of $L$. Assuming that $J$ is exponentially distributed (following the empirical analysis in Sect. 3.3), we find that

$$\psi_J(x) = \ln\left(\frac{a}{ix - a}\right).$$

Here, $a > 0$ is the parameter of the exponential distribution, that is, the expectation of $J$ is $1/a$. Then

$$\psi(s, t, x) = \frac{2a - ix}{ix - a} \int_s^t \lambda(u)\, du,$$

and thus,

$$\psi_{R(t)}(x) = \int_0^t \frac{2a - ixg(u)}{ixg(u) - a} \lambda(u)\, du, \qquad (8.20)$$

for the precipitation process. This cumulant function becomes useful when pricing derivatives, a topic which we present in the next Section.

## 8.2    Pricing derivatives on precipitation

We want to price various derivatives written on precipitation. Recall from Subsect. 1.2.3, that at the CME one can trade in rain and snowfall futures for the accumulated amount of precipitation over a month measured in various cities in the US. Furthermore, one can trade in call and put options on these. There are also many contracts traded bilaterally with different payoff structures.

If we let $[\tau_1, \tau_2]$ be the measurement period for a precipitation derivative, the underlying index will then be

$$\text{Ind}(\tau_1, \tau_2) = R(\tau_2) - R(\tau_1) = \int_{\tau_1}^{\tau_2} g(s)\, dL(s). \qquad (8.21)$$

We are interested in pricing derivatives on $\text{Ind}(\tau_1, \tau_2)$ with payoff $f(\text{Ind}(\tau_1, \tau_2))$ at exercise time $\tau_2$. Following the ideas discussed earlier in this monograph, we define the price to be

$$c(t) = e^{-r(\tau_2 - t)} \mathbb{E}_Q\left[f(\text{Ind}(\tau_1, \tau_2)) \,|\, \mathcal{F}_t\right] \qquad (8.22)$$

for $t \leq \tau_1$. Hence, we assume that the option is traded for the period up to the beginning of the measurement period, and that settlement of the contract takes place immediately at the end of the measurement $\tau_2$. We price under a measure $Q$ determined via an Esscher transform. Esscher transform of the probability $P$ is the topic of the next Subsection.

### 8.2.1 The Esscher transform for independent increment processes

The Esscher transform for independent increment processes is introduced and discussed in detail in [Benth, Šaltytė Benth and Koekebakker (2008)]. We present the basic theory for the special case of a driftless and increasing pure jump independent increment process $L$.

Let $\theta : \mathbb{R}_+ \mapsto \mathbb{R}$ be a bounded and measurable function, and define the stochastic process

$$Z(t) = \exp\left(\int_0^t \theta(s)\, dL(s) - \psi(0, t, -i\theta(\cdot))\right), \tag{8.23}$$

for $t \leq \tau$. Here, we let $\tau$ to be some finite time horizon for the market, sufficiently large to include all the measurement periods $[\tau_1, \tau_2]$ of interest. Note that in the definition of $Z(t)$, we have the term $\psi(0, t, -i\theta(\cdot))$, defined as

$$\psi(0, t, -i\theta(\cdot)) = \int_0^t \int_{\mathbb{R}_+} (e^{i(-i\theta(s))z} - 1)\, \ell(dz, ds)$$

$$= \int_0^t \int_{\mathbb{R}_+} (e^{z\theta(s)} - 1)\, \ell(dz, ds)$$

$$= \ln \mathbb{E}\left[\exp\left(\int_0^t \theta(s)\, dL(s)\right)\right],$$

which is well-defined if exponential moments of $L$ exist. Let $k = \sup_{0 \leq t \leq \tau} |\theta(s)|$, and note that

$$\int_0^t \int_{\mathbb{R}_+} (e^{z\theta(s)} - 1)\, \ell(dz, du) \leq \int_0^t \int_{\mathbb{R}_+} (e^{zk} - 1)\, \ell(dz, du) = \ln \mathbb{E}\left[\exp(kL(t))\right].$$

We make the following condition on the existence of exponential moments of $L$.

**Exponential moment condition.** There exists a constant $k > 0$ such that $\mathbb{E}\left[\exp(kL(\tau))\right] < \infty$.

We remark that for a given process $L$, we know the constant $k$, which restricts the flexibility in choosing the function $\theta$ in the definition of the process $Z(t)$ in (8.23).

We observe that for $\theta$ bounded by $k$, the non-negative process $Z(t)$ is well-defined, and has an expectation equal to one. Moreover, for $s \leq t$, we find that

$$\mathbb{E}\left[Z(t) \mid \mathcal{F}_s\right] = \exp(-\psi(0,t,-i\theta(\cdot)))\mathbb{E}\left[\exp\left(\int_0^t \theta(u)\,dL(u)\right) \mid \mathcal{F}_s\right]$$

$$= \exp(-\psi(0,t,-i\theta(\cdot)))\exp\left(\int_0^s \theta(u)\,dL(u)\right)$$

$$\times \mathbb{E}\left[\exp\left(\int_s^t \theta(u)\,dL(u)\right)\right]$$

$$= \exp(-\psi(0,t,-i\theta(\cdot)))\exp\left(\int_0^s \theta(u)\,dL(u)\right)$$

$$\times \exp(\psi(s,t,-i\theta(\cdot)))$$

$$= Z(s),$$

where the independent increment property of $L$ and the $\mathcal{F}_s$-measurability of $\int_0^s \theta(u)\,dL(u)$ were used. Hence, $Z(t)$ is a martingale.

Introduce now the equivalent probability measure $Q \sim P$ with the density process

$$\left.\frac{dQ}{dP}\right|_{\mathcal{F}_t} = Z(t), \qquad (8.24)$$

for $t \leq \tau$. The probability $Q$ is parametrized by the function $\theta$, modelling the *market price of risk*. We call this change of probability measure for the *Esscher transform*. As the next proposition shows, $L$ remains an independent increment process under $Q$.

**Proposition 8.3.** *Under the probability $Q$, $L$ is an independent increment process with cumulant function*

$$\psi_Q(s,t,x) = \psi(s,t,x - i\theta(\cdot)) - \psi(s,t,-i\theta(\cdot)),$$

*for $0 \leq s \leq t \leq \tau$ and $x \in \mathbb{R}$.*

**Proof.**   Recall the cumulant function $\psi$ in (8.4) of the independent increment process $L$ (with zero-drift by assumption on $L$). We compute the cumulant for $L$ under the probability $Q$. Let $0 \leq s \leq t$ and $x \in \mathbb{R}$. Then,

from the definition of the probability $Q$ and Bayes' Formula, we find that
$$\mathbb{E}_Q \left[ \exp \left( ix(L(t) - L(s)) \right) \mid \mathcal{F}_s \right]$$
$$= \mathbb{E} \left[ \exp \left( ix(L(t) - L(s)) \right) \frac{Z(t)}{Z(s)} \mid \mathcal{F}_s \right]$$
$$= \mathbb{E} \left[ \exp \left( \int_s^t (\theta(u) + ix) \, dL(u) \right) \mid \mathcal{F}_s \right]$$
$$\times \exp \left( \psi(0, s, -i\theta(\cdot)) - \psi(0, t, -i\theta(\cdot)) \right)$$
$$= \mathbb{E} \left[ \exp \left( \int_s^t (\theta(u) + ix) \, dL(u) \right) \right] \exp \left( -\psi(s, t, -i\theta(\cdot)) \right).$$
In the last equality the independent increment property of $L$ was used. This shows that $L$ has independent increments under $Q$. Then we have that
$$\psi_Q(s, t, x) = \ln \mathbb{E}_Q \left[ \exp \left( ix(L(t) - L(s)) \right) \right]$$
$$= \psi(s, t, x - i\theta(\cdot)) - \psi(s, t, -i\theta(\cdot)),$$
which proves the Proposition. $\qquad \square$

From the knowledge of the cumulant $\psi$ in (8.4) the compensator measure and drift of $L$ under $Q$ can be computed. We have
$$\psi_Q(s, t, x) = \psi(s, t, x - i\theta(\cdot)) - \psi(s, t, -i\theta(\cdot))$$
$$= \int_s^t \int_{\mathbb{R}_+} \left( e^{i(x - i\theta(u))z} - 1 \right) \ell(dz, du)$$
$$- \int_s^t \int_{\mathbb{R}_+} \left( e^{i(-i\theta(u))z} - 1 \right) \ell(dz, du)$$
$$= \int_s^t \int_{\mathbb{R}_+} \left( e^{ixz} - 1 \right) e^{\theta(u)z} \ell(dz, du).$$
Hence, with respect to the probability $Q$, the independent increment process $L$ will still be driftless but have a new compensator measure, given by
$$\ell_Q(dz, du) = e^{\theta(u)z} \ell(dz, du). \tag{8.25}$$
The compensator measure is exponentially tilted according to the market price of risk. If $\theta(u) > 0$, we get more emphasis on big jumps, while $\theta(u) < 0$ scales the jumps down. Note that if $L$ is a Lévy process, that is, an independent increment process with stationary increments, then $L$ is not a Lévy process under $Q$ unless $\theta$ is a constant. For a constant $\theta$, the Lévy measure becomes $\ell_Q(dz) = \exp(\theta z) \ell(dz)$.

It is simple to see that the cumulant of $R(t)$ under $Q$ is
$$\psi_{R(t), Q}(x) = \ln \mathbb{E} \left[ e^{ix \int_0^t g(s) \, dL(s)} \right] = \psi_Q(0, t, xg(\cdot)). \tag{8.26}$$
We are now ready to price derivatives on precipitation.

### 8.2.2    Pricing

To price derivatives on precipitation, we apply Fourier-based techniques. Assume that $h \in L^1(\mathbb{R})$, where $L^1(\mathbb{R})$ is the space of integrable functions on $\mathbb{R}$. We define the Fourier transform of $h$ to be

$$\widehat{h}(y) = \int_{\mathbb{R}} h(x) e^{-ixy} \, dx. \tag{8.27}$$

If, in addition, $\widehat{h} \in L^1(\mathbb{R})$, then the inverse Fourier transform is given by

$$h(x) = \frac{1}{2\pi} \int_{\mathbb{R}} \widehat{h}(y) e^{ixy} \, dy. \tag{8.28}$$

These definitions are taken from [Folland (1984)]. Note that the sign in the exponential function above may differ from the traditional definition of the Fourier transform.

We want to price derivatives with payoff $f(\mathrm{Ind}(\tau_1, \tau_2))$ at time $\tau_2$. The approach uses the Fourier transform of $f$, but this is not always an integrable function for natural choices of $f$. For example, considering a call option with strike $K$ will yield the function $f(x) = \max(x - K, 0)$, which is not an element of $L^1(\mathbb{R})$. On the other hand, we can always dampen such function with an exponential one. Hence, we assume that there exists a $\delta > 0$ such that the function $x \mapsto \exp(-\delta x) f(x)$ is integrable. For simplicity, we denote this function by $f_\delta(x)$. We have the following general result on pricing precipitation derivatives.

**Proposition 8.4.** *Assume that $\widehat{f}_\delta \in L^1(\mathbb{R})$ where $\widehat{f}_\delta$ is the Fourier transform of $f_\delta$ and $\sup_{u \in [\tau_1, \tau_2]}(\delta|g(u)| + |\theta(u)|) \leq k$ in the exponential moment condition of $L$. Then, the price $c(t)$ at time $t \leq \tau_1$ of a derivative with payoff $f(Ind(\tau_1, \tau_2))$ at time $\tau_2$ is*

$$c(t) = e^{-r(\tau_2 - t)} \frac{1}{2\pi} \int_{\mathbb{R}} \widehat{f}_\delta(y) \exp\left(\Psi(\tau_1, \tau_2, g(\cdot), \theta(\cdot), y)\right) \, dy,$$

*where*

$$\Psi(\tau_1, \tau_2, g(\cdot), \theta(\cdot), y) = \psi(\tau_1, \tau_2, yg(\cdot) - i(\delta g(\cdot) + \theta(\cdot))) - \psi(\tau_1, \tau_2, -i\theta(\cdot)).$$

**Proof.** We compute the expectation $\mathbb{E}_Q[f(\mathrm{Ind}(\tau_1, \tau_2)) \mid \mathcal{F}_t]$. Since $L$ is an independent increment process under $Q$, by definition of $R(t)$, $\mathrm{Ind}(\tau_1, \tau_2) = R(\tau_2) - R(\tau_1) = \int_{\tau_1}^{\tau_2} g(s) \, dL(s)$ is independent of $\mathcal{F}_t$. Hence,

$$\mathbb{E}_Q\left[f(\mathrm{Ind}(\tau_1, \tau_2)) \mid \mathcal{F}_t\right] = \mathbb{E}_Q\left[f(\mathrm{Ind}(\tau_1, \tau_2))\right].$$

As, by assumption, we have

$$f(x) = \frac{1}{2\pi} \int_{\mathbb{R}} \widehat{f}_\delta(y) e^{(\delta + iy)x} \, dy,$$

it holds that

$$\mathbb{E}_Q\left[f(\mathrm{Ind}(\tau_1,\tau_2))\right] = \frac{1}{2\pi}\int_{\mathbb{R}}\widehat{f}_\delta(y)\mathbb{E}_Q\left[e^{(\delta+iy)\mathrm{Ind}(\tau_1,\tau_2)}\right]\,dy$$

by applying the Fubini-Tonelli Theorem (see [Folland (1984)]). Because $\mathrm{Ind}(\tau_1,\tau_2) = R(\tau_2) - R(\tau_1)$, the cumulant becomes

$$\begin{aligned}
\psi_{R(\tau_2)-R(\tau_1),Q}(x) &= \psi_{R(\tau_2),Q}(x) - \psi_{R(\tau_1),Q}(x)\\
&= \psi_Q(\tau_1,\tau_2,xg(\cdot))\\
&= \psi(\tau_1,\tau_2,xg(\cdot)-i\theta(\cdot)) - \psi(\tau_1,\tau_2,-i\theta(\cdot))\,.
\end{aligned}$$

Therefore

$$\begin{aligned}
\ln\mathbb{E}_Q&\left[e^{(\delta+iy)\mathrm{Ind}(\tau_1,\tau_2)}\right]\\
&= \psi(\tau_1,\tau_2,(y-i\delta)g(\cdot)-i\theta(\cdot)) - \psi(\tau_1,\tau_2,-i\theta(\cdot))\\
&= \psi(\tau_1,\tau_2,yg(\cdot)-i(\delta g(\cdot)+\theta(\cdot))) - \psi(\tau_1,\tau_2,-i\theta(\cdot))\,,
\end{aligned}$$

and the proof is complete. $\square$

We observe that the price $c(t)$ is independent of the current state of $R(t)$ due to the model assumption.

Let us consider some examples. First, assume that

$$f(x) = \max(x - K, 0)\,, \tag{8.29}$$

a call option on precipitation with strike $K > 0$. We have the following Lemma.

**Lemma 8.2.** *For $f$ defined in (8.29) and any $\delta > 0$, it holds that*

$$\widehat{f}_\delta(y) = \frac{1}{(\delta+iy)^2}e^{-(\delta+iy)K}\,.$$

*Moreover, $\widehat{f}_\delta \in L^1(\mathbb{R})$.*

**Proof.** First, we find that $\exp(-\delta x)f(x) \leq x\exp(-\delta x)$ for $x \geq K$, and obviously $f_\delta(x) = 0$ for $x < K$. Hence, $f_\delta \in L^1(\mathbb{R})$ for all $\delta > 0$. A direct computation shows the Fourier transform of $f_\delta$.

$$\begin{aligned}
\widehat{f}_\delta(y) &= \int_{\mathbb{R}}e^{-\delta x}\max(x-K,0)e^{-ixy}\,dx\\
&= \int_K^\infty (x-K)e^{-(\delta+iy)x}\,dx\\
&= \int_K^\infty xe^{-(\delta+iy)x}\,dx - K\int_K^\infty e^{-(\delta+iy)x}\,dx\\
&= \frac{1}{(\delta+iy)^2}e^{-(\delta+iy)K}\,,
\end{aligned}$$

where integration-by-parts was applied in the last equality. Finally, as $|\widehat{f_\delta}(y)| = \exp(-\delta K)(\delta^2 + y^2)^{-1}$, we have that $\widehat{f_\delta} \in L^1(\mathbb{R})$. The Lemma follows.     □

Recall now the cumulant function in (8.20) for $R(t)$ in the case of a compound Poisson process specification of $L$. From Prop. 8.4, we find the function $\Psi(\tau_1, \tau_2, g(\cdot), \theta(\cdot), y)$ as

$$\Psi(\tau_1, \tau_2, g(\cdot), \theta(\cdot), y)$$
$$= \psi(\tau_1, \tau_2, yg(\cdot) - i(\delta g(\cdot) + \theta(\cdot))) - \psi(\tau_1, \tau_2, -i\theta(\cdot))$$
$$= \int_{\tau_1}^{\tau_2} \left( \frac{2a - iyg(u) - (\delta g(u) + \theta(u))}{iyg(u) + (\delta g(u) + \theta(u)) - a} - \frac{2a - \theta(u)}{\theta(u) - a} \right) \lambda(u)\, du\,.$$

Knowing $g$ and $\theta$ enables computing the price $c(t)$ in Prop. 8.4 by performing a numerical integration. The expression of $c(t)$ is suitable for applying the fast Fourier transform (FFT). As it is also necessary to integrate over time in the expression of $\Psi(\tau_1, \tau_2, g(\cdot), \theta(\cdot), y)$, it makes the application of FFT more computationally complex.

In the next example, a swap contract is considered. Let $F(t, \tau_1, \tau_2)$ be the swap price for receiving the money equivalent of $\text{Ind}(\tau_1, \tau_2)$ at time $\tau_2$ in return for $F(t, \tau_1, \tau_2)$. Here we assume that the swap is entered costlessly at time $t \leq \tau_1$. Hence, we have

$$\mathbb{E}_Q\left[\text{Ind}(\tau_1, \tau_2) - F(t, \tau_1, \tau_2) \,|\, \mathcal{F}_t\right] = 0$$

or

$$F(t, \tau_1, \tau_2) = \mathbb{E}_Q\left[\text{Ind}(\tau_1, \tau_2) \,|\, \mathcal{F}_t\right],\qquad(8.30)$$

since the swap price $F(t, \tau_1, \tau_2)$ is naturally adapted to $\mathcal{F}_t$. However, $\text{Ind}(\tau_1, \tau_2)$ is independent of $\mathcal{F}_t$, so we find that

$$F(t, \tau_1, \tau_2) = \mathbb{E}\left[\text{Ind}(\tau_1, \tau_2)\right] = -i\psi'_{R(\tau_2)-R(\tau_1),Q}(0)\,.$$

But,

$$\psi'_{R(\tau_2)-R(\tau_1),Q}(0) = \frac{d}{dx} \int_{\tau_1}^{\tau_2} \int_{\mathbb{R}_+} \left( e^{i(xg(u)-i\theta(u))z} - 1 \right) \ell(dz, du)\bigg|_{x=0}$$
$$= \int_{\tau_1}^{\tau_2} \int_{\mathbb{R}_+} ig(u)z e^{i(xg(u)-i\theta(u))z}\, \ell(dz, du)\bigg|_{x=0}$$
$$= i\int_{\tau_1}^{\tau_2} \int_{\mathbb{R}_+} g(u)z e^{\theta(u)z}\, \ell(dz, du)\,.$$

Thus, the swap price is

$$F(t, \tau_1, \tau_2) = \int_{\tau_1}^{\tau_2} \int_{\mathbb{R}_+} g(u)z e^{\theta(u)z}\, \ell(dz, du)\,.\qquad(8.31)$$

Note that the price is independent of $t$, yielding a constant swap price.

A drawback of the proposed model for precipitation $R(t)$ is that a stochastic price *dynamics* for derivatives is not possible. In a marketplace, one would expect some random fluctuations of the prices over time. Obviously, the reason for the lack of dynamics is the application of an independent increment process $L$ in the specification of $R$. One could imagine several extensions of the model that could mend this defiance, and also introduce a more realistic dynamics of the actual precipitation process $R(t)$. A first obvious choice could be to consider stochastic intensities $\lambda(t)$ in the Poisson process. Such processes are known as doubly stochastic Poisson processes, or Cox processes, and use of a positive process $\lambda(t)$. It could be specified as an increasing independent increment process as discussed in this Chapter. We refer the interested reader to [Brémaud (1981)] for an extensive analysis of doubly stochastic Poisson processes. Using this specification, the process $L$ loses its independent increment property (although it will have conditionally independent increments) and the prices depend explicitly on the intensity. Hence, the derivatives would have a price dynamics stochastically varying in time. Note that by choosing $\lambda(t)$ to follow an independent increment process, we open for seasonality variations as it was detected in Lithuanian precipitation data. A much more sophisticated Markovian point process model for rainfall is considered by [Carmona and Diko (2005)]. We discuss this model in Subsect. 9.1.1, where we also analyze a different framework for pricing of precipitation derivatives.

# Chapter 9

# Utility-based approaches to pricing weather derivatives

In Chapter 5, we considered the problem of weather forwards and futures pricing by means of conditional expectation under some pricing measure $Q$ chosen via the Girsanov transform. This way of pricing weather derivatives may be viewed as based on an actuarial approach, where the fair price is given by the expected payment from the derivative and adjusted by a risk loading modelled via the market price of risk $\theta$ in the Girsanov transform. In the considered context, we relied on the arbitrage theory, where any tradeable instrument should be a martingale in the risk-neutral world, otherwise arbitrage possibilities may exist. In a rather liquid market as futures on temperature, say, this seems to be a reasonable approach.

This way of thinking may not be suitable in many situations relevant in weather derivatives pricing, and other approaches have been proposed in the literature. In fact, many weather derivatives contracts do not have any second-hand market, and thus in reality one enters a contract which is impossible to get out of. We present here utility-based pricing methods, taking this into account.

Both the theory and analysis presented in this Chapter rest on stochastic control theory, and we assume some basic knowledge of this. We refer to [Øksendal (1998)] for an introduction to it.

## 9.1 Indifference pricing

Pricing illiquidly traded derivatives by the *indifference method* has gained a lot of attention in the academic literature. Weather contracts based on temperature and precipitation have been priced by such methods, and we refer to [Carmona (2009)] for a general introduction on indifference pricing applied in various areas of finance. In particular, Chapter 7 in

[Carmona (2009)] deals with weather markets. For a very general approach for indifference pricing of temperature derivatives, we refer to [Ankirchner, Imkeller and Dos Reis (2010)]. In this Section we present the case of a derivative written on a weather index, and derive a price which an *issuer* of the derivative will find reasonable given the risk preferences.

Let us suppose that there exists a market consisting of an asset dependent on the weather index, and a risk-free bond. For example, such an asset may be the stock value of a firm which is heavily exposed to weather, like a power producer or a holiday resort. Any investor in this market may optimize the wealth by allocating money in the asset and the bond. But, the investor may also decide to *sell/issue* the weather derivative in addition to investing in the market. The price of the weather index derivative is defined as the increase in the initial wealth which makes the investor *indifferent* between the two investment alternatives, issuing or not issuing the weather derivative.

To make matters more precise, we assume that the asset depending on the temperature index follows a geometric Brownian motion

$$dS(t) = \mu S(t)\, dt + \sigma S(t)\, dB(t)\,. \tag{9.1}$$

The bond has a risk-free rate of return $r > 0$, and we assume that interest is continuously compounded, that is,

$$dD(t) = rD(t)\, dt \text{ and } D(0) = 1\,. \tag{9.2}$$

In order to be able to use the indifference pricing theory efficiently, we model the weather index directly as a stochastic process. Assume that the index follows a geometric Brownian motion as well,

$$dY(t) = \alpha Y(t)\, dt + \beta Y(t)\, dW(t)\,, \tag{9.3}$$

where $W(t)$ is a Brownian motion correlated with $B(t)$, with correlation coefficient $\rho \in (-1, 1)$. It is a usual setup in this theory, and we refer to [Davis (2001)] for a motivation and empirical considerations in the case of temperature. Earlier in this book, any weather index has been defined by measuring a weather factor like temperature, wind speed or precipitation over a specific period $[\tau_1, \tau_2]$. In the current context, we consider a moving window of measurements of the weather factor in question. For example, $Y(t)$ may model the HDD index measured over the period $t - y$ up to $t$, where $y$ is the length of the measurement period. [Davis (2001)] suggests that the HDD index over a month is close to lognormally distributed for data from London, UK, and based on this finds that a geometric Brownian dynamics is a sensible approximation. The payoff at maturity time $\tau$ (also called the exercise time) of the option is $f(Y(\tau))$, for some bounded payoff function $f(x)$. For example, this may be a put option, in which case $f(x) = \max(K - x, 0)$, where the strike price is $K > 0$.

Let now $(a(t), b(t))$ be the number of stocks and bonds, respectively, in the portfolio of the investor. Hence, the wealth at time $t$ is

$$X(t) = a(t)S(t) + b(t)D(t) \, . \tag{9.4}$$

If we assume that no external funding is injected into the portfolio, or no proceeds are withdrawn, we call it *self-financing* and observe that in this case any change in value of the investor's wealth must come from the change in value of the stock and/or bond. Mathematically, the fact that $X(t)$ is self-financing means

$$dX(t) = a(t) \, dS(t) + b(t) \, dD(t) \, .$$

Now, using that $b(t) = (X(t) - a(t)S(t))/D(t)$, and recalling the dynamics of $S$ and $D$, we find

$$dX(t) = ((\mu - r)a(t)S(t) + rX(t)) \, dt + \sigma a(t)S(t) \, dB(t) \, .$$

Introduce now $\pi(t) = a(t)S(t)$, being the *cash amount* invested in the stock. Then we have the wealth process

$$dX(t) = ((\mu - r)\pi(t) + rX(t)) \, dt + \sigma \pi(t) \, dB(t) \, . \tag{9.5}$$

The investor controls the performance of the wealth through the cash amount $\pi(t)$ allocated to the stocks.

In order for $X(t)$ in (9.5) to make sense, we need some conditions on the controls $\pi(t)$. To make the Itô integral in the dynamics of $X(t)$ well-defined, we impose the condition that $\pi(t)$ is an $\mathcal{F}_t$-adapted stochastic process satisfying

$$\mathbb{E}\left[ \int_0^\tau \pi^2(t) \, dt \right] < \infty \, ,$$

where we recall the time $\tau < \infty$ to be the exercise time of the option. The conditions turn $\pi(t)$ into an Itô integrable process, see e.g. [Øksendal (1998)]. Note that the adaptedness conditon on $\pi(t)$ means that investment decisions can only be made based on the past knowledge collected in $\mathcal{F}_t$, but not on the future. This seems reasonable from a practical point of view, since any precise information about the future is not accessible. Finally, we assume that all strategies are dependent on current time $t$, state of the wealth $X(t)$ and possibly the state of the index $Y(t)$, that is, $\pi(t) := \pi(t, X(t), Y(t))$. Such controls are called *Markovian*.

We assume that the investor has risk preferences given by an exponential utility function, that is, a utility function given by

$$u(x) = -\exp(-\gamma x),$$

with $\gamma > 0$ measuring the risk aversion. The investor then faces the stochastic control problem

$$U(t, x) = \sup_{\pi} \mathbb{E}\left[-\exp\left(-\gamma X(\tau)\right) \mid X(t) = x\right], \qquad (9.6)$$

if no weather index derivative is issued. Controls are naturally depending only on the state of the wealth and not on the index in this case. On the other hand, if the investor issues such a derivative, the financial position is then optimized by solving the stochastic control problem

$$V(t, x, y) = \sup_{\pi} \mathbb{E}\left[-\exp\left(-\gamma(X(\tau) - f(Y(\tau)))\right) \mid X(t) = x, Y(t) = y\right]. \qquad (9.7)$$

In order for the investor to be indifferent between the two situations, an initial premium $p(t, x, y)$ for issuing the derivative, defined as the solution to the equation

$$V(t, x + p(t, x, y), y) = U(t, x), \qquad (9.8)$$

should be charged. If a function $p(t, x, y)$ solving equation (9.8) exists, we call this function the *indifference price*. It will be the price that makes the investor indifferent between issuing or not issuing the derivative.

The aim of the rest of this Section is to derive an expression for $p(t, x, y)$ based on stochastic control theory. We will make use of the traditional dynamic programming approach. The interested reader will find an introduction to this in [Øksendal (1998)]. We first solve the optimal portfolio problem without any weather derivative issued. From dynamic programming, one can derive the Hamilton-Jacobi-Bellman (HJB) equation for $U(t, x)$ in (9.6) as

$$\partial_t U + \max_{\pi}\left(((\mu - r)\pi + rx)\partial_x U + \frac{1}{2}\sigma^2\pi^2\partial_{xx}U\right) = 0. \qquad (9.9)$$

Here, $\partial_x U$ and $\partial_t U$ are the derivatives of $U$ with respect to $x$ and $t$, respectively, while $\partial_{xx}U$ is the second-order derivative with respect to $x$.

By differentiating with respect to $\pi$ under the max-operator in (9.9), we get that the first-order condition for optimality is

$$\sigma^2\pi\partial_{xx}U + (\mu - r)\partial_x U = 0$$

or

$$\pi = -\frac{(\mu - r)\partial_x U}{\sigma^2 \partial_{xx} U}. \tag{9.10}$$

Inserting this into the HJB equation (9.9), we get the following nonlinear parabolic partial differential equation for $U$

$$\partial_t U + rx\partial_x U - \frac{(\mu - r)^2}{2\sigma^2}\frac{(\partial_x U)^2}{\partial_{xx} U} = 0. \tag{9.11}$$

In addition, from (9.6) we find the terminal condition

$$U(\tau, x) = -\exp(-\gamma x). \tag{9.12}$$

We guess a solution of the form

$$U(t, x) = h(t)\exp(-\gamma g(t)x).$$

After finding the differentials, we insert the result into the HJB equation (9.9) to obtain

$$\frac{h'(t)}{h(t)} - \frac{(\mu - r)^2}{2\sigma^2} = (\gamma g'(t) + r\gamma g(t))x.$$

Thus, since $x$ is arbitrary and both sides must be equal, we find that $h$ and $g$ solve the ordinary differential equations

$$h'(t) = \frac{(\mu - r)^2}{2\sigma^2}h(t),$$

$$g'(t) = -r\,g(t),$$

with terminal conditions $g(\tau) = 1$ and $h(\tau) = -1$. We therefore have

$$h(t) = -\exp\left(-\frac{(\mu - r)^2}{2\sigma^2}(\tau - t)\right), \tag{9.13}$$

$$g(t) = \exp(r(\tau - t)). \tag{9.14}$$

In conclusion, the optimal value function of the control problem without any weather derivative issued becomes

$$U(t, x) = -\exp\left(-\frac{(\mu - r)^2}{2\sigma^2}(\tau - t) - \gamma e^{r(\tau - t)}x\right). \tag{9.15}$$

From (9.10), this value is achieved by using the optimal control

$$\pi(t, x) = \frac{\mu - r}{\gamma\sigma^2}e^{-r(\tau - t)}. \tag{9.16}$$

Hence, the cash amount invested in the risky asset at time $t$ should be the discounted value of the risk-adjusted return over volatility. If $\tau - t$ is large, that is, we are far from maturity of the derivative, we should hold close to nothing in the stock, whereas close to maturity the cash invested in asset should be $(\mu - r)/\gamma\sigma^2$.

By applying a so-called verification theorem, one may in a mathematical rigorous way prove that $U(t, x)$ and $\pi(t, x)$ in (9.15) and (9.16), respectively, are solutions to the stochastic control problem. We refer to [Øksendal (1998)] for more mathematical details.

We continue with the case where the investor issues a weather derivative. The HJB equation for $V(t, x, y)$ defined in (9.7) is

$$\partial_t V + \max_\pi \left( ((\mu - r)\pi + rx) \, \partial_x V + \frac{1}{2}\sigma^2\pi^2\partial_{xx}V + \rho\sigma\beta\pi y\partial_{xy}V \right)$$
$$+ \alpha y\partial_y V + \frac{1}{2}\beta^2 y^2\partial_{yy}V = 0. \tag{9.17}$$

The first-order condition for optimality is

$$(\mu - r)\partial_x V + \sigma^2\pi\partial_{xx}V + \rho\sigma\beta y\partial_{xy}V = 0,$$

which, after solving for $\pi$, becomes

$$\pi = -\frac{(\mu - r)\partial_x V + \rho\sigma\beta y\partial_{xy}V}{\sigma^2\partial_{xx}V}. \tag{9.18}$$

Inserting this into (9.17), we obtain the nonlinear partial differential equation

$$\partial_t V + rx\partial_x V - \frac{((\mu - r)\partial_x V + \rho\sigma\beta y\partial_{xy}V)^2}{2\sigma^2\partial_{xx}V}$$
$$+ \rho\frac{\beta}{\sigma}(\mu - r)y\frac{\partial_{xy}V\partial_x V}{\partial_{xx}V} + \alpha y\partial_y V + \frac{1}{2}\beta^2 y^2\partial_{yy}V = 0, \tag{9.19}$$

with the terminal value derived from (9.7) as

$$V(\tau, x, y) = -\exp(-\gamma(x - f(y))). \tag{9.20}$$

We now solve this nonlinear partial differential equation.

The nonlinear partial differential equation for $V$ in (9.19) is considerably more complex than for $U(t, x)$. [Musiela and Zariphopoulou (2004)] used an ingenious power transform to express the solution as

$$V(t, x, y) = U(t, x)F^\delta(t, y),$$

for some smooth function $F$ and a constant $\delta > 0$. Indeed, from the terminal condition, we get

$$V(\tau, x, y) = U(\tau, x)F^\delta(\tau, y) = -\exp(-\gamma x)\exp(\gamma f(y))$$

or

$$F(\tau, y) = \exp\left(\frac{\gamma}{\delta} f(y)\right).$$

Finding the derivatives of $V$ and inserting into (9.19), we obtain

$$\partial_t U F^\delta + U \delta F^{\delta-1} \partial_t F + rx \partial_x U F^\delta$$

$$- \frac{\left((\mu - r)\partial_x U F^\delta + \rho\sigma\beta y \partial_x U \delta F^{\delta-1} \partial_y F\right)^2}{2\sigma^2 \partial_{xx} U F^\delta}$$

$$+ \rho\frac{\beta}{\sigma}(\mu - r)y \frac{\partial_x U \delta F^{\delta-1} \partial_y F \partial_x U F^\delta}{\partial_{xx} U F^\delta} + \alpha y U \delta F^{\delta-1} \partial_y F$$

$$+ \frac{1}{2}\beta^2 y^2 U \left(\delta(\delta - 1) F^{\delta-2}(\partial_x F)^2 + \delta F^{\delta-1}\partial_{yy}F\right) = 0.$$

Dividing by $F^\delta$ and reorganizing give

$$\left(\partial_t U + rx\partial_x U - \frac{(\mu - r)^2(\partial_x U)^2}{2\sigma^2 \partial_{xx}U}\right)$$

$$+ \delta\frac{U}{F}\left(\partial_t F + \alpha y \partial_y F + \frac{1}{2}\beta^2 y^2 \partial_{yy}F\right)$$

$$+ \frac{1}{2}\beta^2 y^2 \left(\frac{\partial_y F}{F}\right)^2 \left(\delta(\delta - 1)U - \rho^2\delta^2\frac{(\partial_x U)^2}{\partial_{xx}U}\right) = 0.$$

The first term is zero because $U$ is a solution of (9.11). Furthermore, since

$$\frac{(\partial_x U)^2}{\partial_{xx}U} = U$$

we find, after choosing

$$\delta = \frac{1}{1 - \rho^2},$$

the linear partial differential equation of $F$,

$$\partial_t F + \alpha y \partial_y F + \frac{1}{2}\beta^2 y^2 \partial_{yy}F = 0, \tag{9.21}$$

with initial condition

$$F(\tau, y) = \exp\left(\gamma(1 - \rho^2)f(y)\right). \tag{9.22}$$

We recognize that this *linear* partial differential equation has the same form as the partial differential equation for the Black-Scholes call option price (see [Benth (2004)]).

The solution of (9.21) can be represented as an expectation by (see [Øksendal (1998)])

$$F(t, y) = \mathbb{E}\left[\exp\left(\gamma(1 - \rho^2)f(Y(T))\right) \mid Y(t) = y\right]. \tag{9.23}$$

It is simple to see that the distribution of the lognormal random variable $Y(T)$, given $Y(t) = y$, is the fundamental solution of the partial differential equation (9.21), and thus the expression for $F(t, y)$ in (9.23) is just a stochastic way to represent the solution of (9.21) that could be derived by Fourier transform. Frequently, one refers to (9.23) as the Feynman-Kac solution of (9.21).

Collecting our findings for the optimal portfolio problem with a weather derivative issued, we conclude that

$$V(t, x, y) = U(t, x) \mathbb{E} \left[ \exp \left( \gamma (1 - \rho^2) f(Y(T)) \right) \mid Y(t) = y \right]^{\frac{1}{1 - \rho^2}} . \qquad (9.24)$$

With this expression at hand, we can easily compute the indifference price.

**Proposition 9.1.** *The indifference price $p(t, x, y)$ defined in (9.8) of the weather derivative with payoff $f(Y(\tau))$ is given by*

$$p(t, x, y) = \frac{1}{\gamma(1 - \rho^2)} e^{-r(\tau - t)} \ln \mathbb{E} \left[ \exp \left( \gamma (1 - \rho^2) f(Y(\tau)) \right) \mid Y(t) = y \right] .$$

**Proof.** From the definition of the indifference price as the solution of equation (9.8), we have

$$U(t, x + p(t, x, y)) F(t, y)^{1/1 - \rho^2} = U(t, x) ,$$

for $U$ given in (9.15) and $F$ in (9.23). Since

$$\frac{U(t, x + p)}{U(t, x)} = \exp \left( -\gamma e^{r(\tau - t)} p \right) ,$$

the result follows. $\qquad \square$

Let us derive the optimal control for the investor who has issued the derivative. Since

$$\partial_x V = \partial_x U \, F^\delta , \quad \partial_{xx} V = \partial_{xx} U \, F^\delta , \quad \partial_{xy} V = \delta \partial_x U \, F^{\delta - 1} \partial_y F$$

with $F$ as in (9.23) and $\delta = 1/(1 - \rho^2)$, we find the optimal investment strategy $\pi$ defined in (9.18) as

$$\pi(t, x, y) = -\frac{\mu - r}{\sigma^2} \frac{\partial_x U}{\partial_{xx} U} - \rho \delta \frac{\beta}{\sigma} y \frac{\partial_x U}{\partial_{xx} U} \frac{\partial_y F}{F} .$$

But, $\partial_x U / \partial_{xx} U = -(1/\gamma) \exp(-r(\tau - t))$, and therefore, since $\delta = 1/(1 - \rho^2)$, we have

$$\pi(t, x, y) = \frac{\mu - r}{\gamma \sigma^2} e^{-r(\tau - t)} + \frac{\rho}{1 - \rho^2} \frac{\beta}{\gamma \sigma} y e^{-r(\tau - t)} \frac{\partial_y F(t, y)}{F(t, y)} . \qquad (9.25)$$

This strategy is not dependent on the current level of wealth $x$, but on the index value $y$. We observe that the first term coincides with the optimal investment strategy if no derivative is issued (recall (9.25)).

We may interpret the difference between the optimal investment strategy with a derivative issued, and the one without, as the *hedging strategy* of the derivative. This will not be a perfect hedge, but it provides the "optimal" protection against the risk in issuing the weather derivative and pricing it by the indifference method. We next find the hedging strategy.

**Proposition 9.2.** *The optimal hedging strategy from indifference pricing is given by*

$$\pi_h(t, y) = \frac{\rho}{1 - \rho^2} \frac{\beta}{\gamma\sigma} y e^{-r(\tau - t)} \frac{\partial_y F(t, y)}{F(t, y)} ,$$

*where $F(t, y)$ is defined in (9.23).*

**Proof.** This is immediately seen by taking the difference between the optimal strategy in (9.16) and the optimal strategy in (9.25).     □

We next investigate the ratio $\partial_y F / F$ appearing in the hedging strategy. From Itô's Formula in Thm. 4.1 we find that $Y(\tau)$ conditioned on $Y(t) = y$, $\tau \geq t$, is equal to

$$Y^{t,y}(\tau) = y \exp\left(\left(\alpha - \frac{1}{2}\beta^2\right)(\tau - t) + \beta(W(\tau) - W(t))\right) .$$

Here $Y^{t,y}(t) = y$. Hence

$$F(t, y) = \mathbb{E}\left[\exp\left(\gamma(1 - \rho^2)f(Y^{t,y}(\tau))\right)\right] .$$

If $f$ is a differentiable function, with a bounded derivative, we find by applying the dominated convergence (see [Folland (1984)]) that

$$\partial_y F(t, y) = \gamma(1 - \rho^2)\mathbb{E}\left[f'(Y^{t,y}(\tau))Y^{t,1}(\tau)\exp\left(\gamma(1 - \rho^2)f(Y^{t,y}(\tau))\right)\right] ,$$

since $\partial_y Y^{t,y}(\tau) = Y^{t,1}(\tau)$. Hence, the hedging strategy becomes

$$\pi_h(t, y) = \rho\frac{\beta}{\sigma} y e^{-r(\tau - t)} \frac{\mathbb{E}\left[f'(Y^{t,y}(\tau))Y^{t,1}(\tau)\exp\left(\gamma(1 - \rho^2)f(Y^{t,y}(\tau))\right)\right]}{\mathbb{E}\left[\exp\left(\gamma(1 - \rho^2)f(Y^{t,y}(\tau))\right)\right]} .$$
$$(9.26)$$

Observe the factor $\rho\beta/\sigma$, telling that the hedging position in the risky asset $X$ is proportional to the correlation $\rho$ with the weather index times the ratio between the volatility of the weather index and the risky asset.

One may represent the derivative of $F(t, y)$ with respect to $y$ without having to differentiate the payoff function $f$. If we are facing the problem

of pricing a digital option on the weather index, for example, an option
paying one unit of a currency in case the index is higher than a prescribed
threshold but zero otherwise, the derivative of the payoff function is zero
almost everywhere (the derivative is not defined at the threshold). We can
thus represent $Y^{t,y}(\tau)$ as

$$Y^{t,y}(\tau) = \exp\left(\ln(y) + \left(\alpha - \frac{1}{2}\beta^2\right)(\tau - t) + \beta\sqrt{\tau - t}Z\right),$$

where equality is in distribution and $Z$ is a standard normally distributed
random variable. If $\phi(z)$ denotes the density function of a standard normal
random variable, we find that

$$F(t,y) = \int_{\mathbb{R}} e^{\gamma(1-\rho^2)f(\exp(x))}\frac{1}{\beta\sqrt{\tau - t}}\phi\left(\frac{x - (\ln(y) + (\alpha - \frac{1}{2}\beta^2)(\tau - t))}{\beta\sqrt{\tau - t}}\right)dx.$$

Then, after commuting differentiation and integration, and using the fact
that $\phi'(z) = -z\phi(z)$, we get

$$\partial_y F(t,y)$$

$$= \int_{\mathbb{R}} e^{\gamma(1-\rho^2)f(\exp(x))}\frac{1}{\beta\sqrt{\tau - t}}\phi'\left(\frac{x - (\ln(y) + (\alpha - \frac{1}{2}\beta^2)(\tau - t))}{\beta\sqrt{\tau - t}}\right)$$

$$\times \frac{-1}{\beta\sqrt{\tau - t}}\frac{1}{y}\,dx$$

$$= \frac{1}{y}\frac{1}{\beta^2(\tau - t)}\mathbb{E}\left[\left(\ln(Y^{t,y}(\tau)) - \ln(y) - (\alpha - \frac{1}{2}\beta^2)(\tau - t)\right)\right.$$

$$\left. \times \exp\left(\gamma(1 - \rho^2)f(Y^{t,y}(\tau))\right)\right].$$

This expression for $\partial_y F(t,y)$ yields a formula for the optimal hedge which
does not involve the derivative of the payoff function $f$. We remark that
one often refers to the above approach for computing derivative-free hedges
as the *density method* (see [Glasserman (2003)] for more on this method).

Let us discuss some properties of the indifference price in Prop. 9.1. It
is immediately seen that the price does not depend on the current level of
wealth $x$, but only time $t$ and current value $y$ of the index. Obviously, the
indifference price is a function of the risk aversion parameter $\gamma$. A simple
application of L'Hopital's rule reveals that

$$\lim_{\gamma\downarrow 0} p(t,x,y) = e^{-r(\tau - t)}\mathbb{E}\left[f(Y(\tau))\,|\,Y(t) = y\right].$$

Hence, an investor who is "risk indifferent" will price the derivative using
the expected present value of the payoff. Note that there is no risk adjust-
ment by a pricing measure $Q$, but the expected value is under the market
probability $P$. We find from Jensen's inequality (see [Folland (1984)]) that

$$p(t,x,y) \geq e^{-r(\tau - t)}\mathbb{E}\left[f(Y(\tau))\,|\,Y(t) = y\right].$$

Hence, the indifference price for risk aversion coefficients $\gamma > 0$ will be greater than the benchmark case $\gamma = 0$. This is natural from the point of view that a risk averse investor should charge more for taking on the risk of issuing a derivative than an investor which is neutral towards risk. The following proposition shows that the indifference price is monotonely increasing as a function of the risk aversion $\gamma$, which is according to what is intuitively expected.

**Proposition 9.3.** *For fixed $(t, x, y)$, the function $\gamma \mapsto p(t, x, y)$ for $p(t, x, y)$ defined in Prop. 9.1 is monotonely increasing in $\gamma \geq 0$.*

**Proof.** Without loss of generality, we assume $r = \rho = 0$, and for notational simplicity we denote the random variable $f(Y^{t,y}(\tau))$ by $G$. We have

$$p(\gamma) := p(t, x, y) = \frac{1}{\gamma} \ln \mathbb{E}[e^{\gamma G}].$$

Here $\psi(\gamma) := \ln \mathbb{E}[\exp(\gamma G)]$ is the logarithm of the moment generating function of $G$. A direct differentiation yields

$$p'(\gamma) = -\frac{1}{\gamma^2} \ln \mathbb{E}[e^{\gamma G}] + \frac{1}{\gamma} \frac{\mathbb{E}[Ge^{\gamma G}]}{\mathbb{E}[e^{\gamma G}]} = \frac{1}{\gamma} \left( \frac{\mathbb{E}[Ge^{\gamma G}]}{\mathbb{E}[e^{\gamma G}]} - p(\gamma) \right).$$

Then $p(\gamma)$ will be increasing as long as the expression within the brackets is non-negative. Since $\psi'(\gamma) = \mathbb{E}[G \exp(\gamma G)]/\mathbb{E}[\exp(\gamma G)]$, this is true if and only if

$$\psi'(\gamma) \geq \frac{1}{\gamma} \psi(\gamma).$$

Define the function

$$g(\gamma) = \psi(\gamma) - \gamma \psi'(\gamma),$$

and observe that $g(0) = \psi(0) - 0 \cdot \psi'(0) = 0$. Obviously,

$$g'(\gamma) = \psi'(\gamma) - \psi'(\gamma) - \gamma \psi''(\gamma) = -\gamma \psi''(\gamma).$$

We compute $\psi''(\gamma)$ to be

$$\psi''(\gamma) = \frac{d}{d\gamma} \frac{\mathbb{E}[Ge^{\gamma G}]}{\mathbb{E}[e^{\gamma G}]} = \frac{\mathbb{E}[G^2 e^{\gamma G}]\mathbb{E}[e^{\gamma G}] - \mathbb{E}[Ge^{\gamma G}]^2}{\mathbb{E}[e^{\gamma G}]^2}.$$

By the Cauchy-Schwartz inequality (see [Folland (1984)]) it follows that

$$\mathbb{E}[Ge^{\gamma G}]^2 = \mathbb{E}[Ge^{\gamma G/2}e^{\gamma G/2}]^2 \leq \mathbb{E}[G^2 e^{\gamma G}]\mathbb{E}[e^{\gamma G}].$$

Thus, the function $g(\gamma)$ is decreasing since $g'(\gamma) = -\gamma \psi''(\gamma) \leq 0$, and we conclude that $p'(\gamma) \geq 0$. The proof is complete. $\square$

Note that the indifference price is nonlinear in terms of the number of contracts issued. In the above the price of one issued weather derivative using the indifference pricing technology is computed. The price of $n$ contracts, denoted $p(t, x, y; n)$ would, by similar derivations, be

$$p(t, x, y; n) = \frac{1}{\gamma(1 - \rho^2)} e^{-r(\tau-t)} \ln \mathbb{E} \left[ \exp\left(\gamma(1 - \rho^2) n f(Y(\tau))\right) \mid Y(t) = y \right].$$
(9.27)

We see that $p(t, x, y; n) \neq np(t, x, y)$. Since $x^n$ is a convex function for an even number $n \geq 2$, Jensen's inequality gives that

$$\mathbb{E} \left[ \left( \exp(\gamma(1 - \rho^2) f(Y(\tau))) \right)^n \right] \geq \mathbb{E} \left[ \exp(\gamma(1 - \rho^2) f(Y(\tau))) \right]^n.$$

Therefore

$$p(t, x, y; n) \geq np(t, x, y).$$

Indeed, this result is a reflection of the risk-averse investor, and therefore a compensation for increasing risk exposure is demanded.

Next, let us briefly discuss the indifference price for the *buyer* of the derivative. Instead of the control problem in (9.7), the buyer will try to maximize

$$\widetilde{V}(t, x, y) = \sup_\pi \mathbb{E} \left[ -\exp\left(-\gamma(X(\tau) + f(Y(\tau)))\right) \mid X(t) = x, Y(t) = y \right].$$
(9.28)

Obviously, buying the derivative adds to the wealth, but for this a price must be paid. The indifference price $q(t, x, y)$ for the buyer is defined to be the solution of the equation

$$\widetilde{V}(t, x - q(t, x, y), y) = U(t, x),$$
(9.29)

with $U$ being the value function in (9.6) when no claim is bought. Doing the same derivations as for the seller's indifference price, we end up with the expression

$$q(t, x, y) = \frac{-1}{\gamma(1 - \rho^2)} e^{-r(\tau-t)} \ln \mathbb{E} \left[ \exp\left(-\gamma(1 - \rho^2) f(Y(\tau))\right) \mid Y(t) = y \right].$$
(9.30)

The buyer's indifference price is also independent of the current wealth $x$.

In the indifference price $q(t, x, y)$, $\gamma$ will, of course, be the risk aversion of the buyer. By taking the limit $\gamma \to 0$, we obtain

$$\lim_{\gamma \downarrow 0} q(t, x, y) = e^{-r(\tau-t)} \mathbb{E} \left[ f(Y(\tau)) \mid Y(t) = y \right],$$

which is the limit of the seller's indifference price. Following the same argument as in the proof of Prop. 9.3, we can show that $q(t, x, y)$ is *decreasing* with $\gamma$. That is, the more risk-averse the buyer is, the lower indifference price will be accepted. This shows that the maximal indifference price for the buyer is attained for $\gamma = 0$, and all other prices are lower than this. In fact, this yields that $q(t, x, y) \leq p(t, x, y)$, as expected. Note that the buyer's and seller's indifference prices only agree when $\gamma = 0$.

In Sect. 9.3 we investigate a slightly different approach motivated by indifference pricing, namely the marginal utility approach advocated by [Davis (2001)] in connection to weather derivatives.

### 9.1.1 *Application to the pricing of rainfall derivatives*

The utility indifference approach has been applied by [Carmona and Diko (2005)] to price rainfall derivatives, using a Markovian model for rainfall intensity. The stochastic process modelling rainfall intensity, denoted $Y(t)$, is a reformulation of the Bartlett-Lewis Poisson-cluster model of [Rodriguez-Iturbe, Cox and Isham (1987)] that we briefly discussed in Subsect. 2.6.3. [Leobacher and Ngare (2011)] propose a related "Markovian gamma" model which accounts for seasonality, and valuate rainfall derivatives by utility indifference pricing.

More precisely, the Markov jump process $Y(t)$ has generator $\mathcal{A}$ defined by

$$\mathcal{A}f(y) = \lim_{t \downarrow 0} \frac{\mathbb{E}[f(Y(t)) \mid Y(0) = y] - f(y)}{t}$$
$$= \int_{-y}^{\infty} (f(y + z) - f(y))A(y, z)\, dz\,, \qquad (9.31)$$

where, for $y \geq 0$,

$$A(y, z) = \lambda_1 \lambda_u e^{-\lambda_u z} \mathbf{1}_{(0, \infty)}(z) + \overline{\lambda}_2(y)\lambda_d e^{\lambda_d z} \mathbf{1}_{(-y, 0)}(z)$$
$$+ \overline{\lambda}_2(y)e^{-\lambda_d y}\mathbf{1}_{\{-y\}}(z)\,, \qquad (9.32)$$

and

$$\overline{\lambda}_2(y) = \lambda_2^{(I)} + \lambda_2^{(II)}\kappa(y)\,. \qquad (9.33)$$

Here, $\lambda_2^{(I)}$ and $\lambda_2^{(II)}$ are two positive constants. The function $\kappa(y)$ is defined on $\mathbb{R}_+$ and assumed to be three times differentiable with positive first derivative. Moreover, it holds that $\kappa(y) = y$ for $0 \leq y \leq K$ for some large $K$. We say that $f$ is in the domain of $\mathcal{A}$ as long as the limit in (9.31) is well-defined pointwise. For theory on Markov jump processes and their generators, see [Ethier and Kurtz (1986)].

Roughly speaking, as seen from the generator $\mathcal{A}$, the process $Y(t)$ changes intensity according to a Poisson process. Each time the rain intensity changes, an exponential increase occurs, either upwards with expected jump size $\lambda_u$, or downwards with expected size $\lambda_d$. However, to preserve a non-negative intensity, we truncate the downward jump by the size of the present state, $y$. Hence, jump downwards appears in the intensity according to a truncated exponential distribution. In addition, we also allow for a jump directly to zero (last term in $A$), corresponding to the state of no rain. Removing the last two terms in the jump size measure $A(y, z)$ (that is, letting $\lambda_2^{(I)} = \lambda_2^{(II)} = 0$), $Y(t)$ becomes a compound Poisson process with intensity $\lambda_1$ and exponentially distributed jumps with expected jump size $\lambda_u$. The measure $A$ is state-dependent, hence the rain intensity is not a Lévy process nor an independent increment process.

Although seemingly complex, the attractiveness of the model $Y(t)$ lies in the possibility to estimate parameters using the maximum likelihood method. This issue is further considered in [Carmona and Diko (2005)].

The process $R(t)$ in Chapter 8 models the accumulated amount of precipitation up to time $t$. In the present context, we can define the accumulated amount of rainfall by $R(t) = \int_0^t Y(t)\,dt$. We obtain a model different than that in Chapter 8 in several ways. First of all, based on $Y(t)$ the cumulative amount of rainfall becomes a smooth process in the sense that it is differentiable. With the model in Chapter 8, we allow for time-dependent jump sizes and intensities, features which are not accounted for in $Y(t)$.

Let us continue with presenting the indifference approach to pricing rainfall derivatives based on the model of [Carmona and Diko (2005)]. Denote $S(t)$ the price at time $t$ of an asset which is dependent on the rainfall intensity $Y(t)$. The dynamics of $S(t)$ is assumed to be of the form

$$dS(t) = S(t)\left(\mu(Y(t))\,dt + \sigma(Y(t))\,dB(t)\right),  \qquad (9.34)$$

that is, a geometric Brownian motion with coefficients depending on the state of the rainfall intensity. The coefficient functions $\mu$ and $\sigma$ are assumed to be sufficiently regular to ensure the existence and uniqueness of $S(t)$ solving the stochastic differential equation (9.34). As in the section above, we let $\pi$ be the cash amount invested in this risky asset, and for simplicity we assume that the interest rate is zero, that is, $r = 0$. Then, for admissible strategies $\pi$, the wealth process becomes

$$dX(t) = \pi(t)\left(\mu(Y(t))\,dt + \sigma(Y(t))\,dB(t)\right).$$

Following the indifference pricing approach, our aim is to solve two stochastic control problems, one classical portfolio optimization problem and another one where there is a derivative on rainfall issued. In both cases we assume that the risk preferences are measured by an exponential utility function.

The utility maximization problem

$$U(t, x, y) = \sup_{\pi} \mathbb{E}\left[-\exp(-\gamma X(\tau)) \mid X(t) = x, Y(t) = y\right],$$

solves, using dynamic programming, the HJB equation

$$\partial_t U + \mathcal{A}U + \max_{\pi}\left(\pi\mu(y)\partial_x U + \frac{1}{2}\pi^2\sigma^2(y)\partial_{xx}U\right) = 0,$$

with terminal condition $U(\tau, x, y) = -\exp(-\gamma x)$. Optimizing with respect to $\pi$ yields

$$\pi = -\frac{\mu(y)\partial_x U}{\sigma^2(y)\partial_{xx}U},$$

which, inserted into the HJB equation, gives an integro-partial differential equation for $U$

$$\partial_t U + \mathcal{A}U - \frac{1}{2}\frac{\mu^2(y)}{\sigma^2(y)}\frac{(\partial_x U)^2}{\partial_{xx}U} = 0. \tag{9.35}$$

By applying a Cole-Hopf transformation, we assume that the value function $U$ is representable as $U(t, x, y) = -\exp(-\gamma x)h(t, y)$ for some sufficiently regular function $h(t, y)$. Substituting into (9.35), we find the integro-differential equation

$$\partial_t h + \mathcal{A}h = \frac{1}{2}\frac{\mu^2(y)}{\sigma^2(y)}h,$$

with terminal condition $h(\tau, y) = 1$. The Feynman-Kac formula[1] says that the solution to this integro-differential equation is

$$h(t, y) = \mathbb{E}\left[\exp\left(-\frac{1}{2}\int_t^\tau \frac{\mu^2(Y(s))}{\sigma^2(Y(s))}\,ds\right) \mid Y(t) = y\right]. \tag{9.36}$$

The optimal investment strategy then becomes

$$\pi(t, y) = \frac{\mu(Y(t-))}{\gamma\sigma^2(Y(t-))},$$

where $Y(t-)$ means the left-limit of the process $Y$, e.g. $\lim_{s \uparrow t} Y(s) = Y(t-)$. We see that the value function and optimal investment strategy are analogous to the results in the section above.

Next, consider the situation where the investor has issued (or sold) a derivative on rainfall. In [Carmona and Diko (2005)], one first restricts to

---

[1]See [Karatzas and Shreve (1991)] for a discussion of the Feynman-Kac formula in the simple Brownian motion case.

derivatives with payoff $\xi = \int_{\tau_1}^{\tau_2} \phi(Y(s)) \, ds$, with a non-negative function $\phi$ of at least polynomial growth. Typically, $\phi(y) = y$, giving a claim on the cumulative amount of rain over the measurement period $[\tau_1, \tau_2]$, or $\phi(y) = \mathbf{1}_{(\epsilon,\infty)}(y)$ giving a claim on the amount of time it rains over the measurement period. The threshold $\epsilon > 0$ indicates the minimal required rainfall to qualify as a "rainy day". The portfolio problem for the seller of the claim becomes

$$V(t, x, y)$$
$$= \sup_{\pi} \mathbb{E} \left[ -\exp\left( -\gamma X(\tau) - \int_{\tau_1}^{\tau_2} \phi(Y(s)) \, ds \right) \,\Big|\, X(t) = x, Y(t) = y \right]$$

for $0 \le \tau_1 < \tau_2 \le \tau$. In Prop. 3.4 of [Carmona and Diko (2005)] it is shown that this optimal portfolio problem is equivalent to the problem

$$\widehat{V}(t, x, y) = \sup_{\pi} \mathbb{E} \left[ -\exp\left( -\gamma \widehat{X}(\tau) \right) \,\Big|\, \widehat{X}(t) = x, Y(t) = y \right] ,$$

where

$$d\widehat{X}(t) = -g(t, Y(t)) \, dt + \pi(t) \left( \mu(Y(t)) \, dt + \sigma(Y(t)) \, dB(t) \right)$$

and

$$g(t, y) = \mathbf{1}_{(\tau_1, \tau_2)}(t) \phi(y) .$$

The HJB equation for $\widehat{V}$ is

$$\partial_t \widehat{V} - g(t, y) \partial_x \widehat{V} + \mathcal{A}V + \max_{\pi} \left( \pi \mu(y) \partial_x \widehat{V} + \frac{1}{2} \pi^2 \sigma^2(y) \partial_{xx} \widehat{V} \right) = 0$$

with terminal condition $\widehat{V}(\tau, x, y) = -\exp(-\gamma x)$. By the same procedure as above, we find the optimal investment strategy as

$$\pi = -\frac{\mu(y)}{\sigma^2(y)} \frac{\partial_x \widehat{V}}{\partial_{xx} \widehat{V}} ,$$

and the corresponding integro-partial differential equation for $\widehat{V}$ as

$$\partial_t \widehat{V} - g(t, y) \partial_x \widehat{V} + \mathcal{A}\widehat{V} - \frac{1}{2} \frac{\mu^2(y)}{\sigma^2(y)} \frac{(\partial_x \widehat{V})^2}{\partial_{xx} \widehat{V}} = 0 .$$

A Cole-Hopf transformation $\widehat{V}(t, x, y) = -\exp(-\gamma x) F(t, y)$ yields the integro-differential equation

$$\partial_t F + \mathcal{A}F = \left( -\gamma g(t, y) + \frac{1}{2} \frac{\mu^2(y)}{\sigma^2(y)} \right) F ,$$

where the solution has the Feynman-Kac representation

$$F(t,y) = \mathbb{E}\left[\exp\left(-\int_t^\tau -\gamma g(s,Y(s)) + \frac{1}{2}\frac{\mu^2(Y(s))}{\sigma^2(Y(s))}\,ds\right)\,\Big|\,Y(t) = y\right].$$

(9.37)

The optimal investment strategy becomes the same as above, $\pi(t,y) = \mu(Y(t-))/(\gamma\sigma^2(Y(t-)))$.

The indifference price $p(t,x,y)$ of the claim $\xi$ for the seller is then given by the solution of the equation

$$V(t, x + p(t,x,y), y) = U(t,x,y),$$

or, equivalently,

$$\exp\left(-\gamma p(t,x,y)\right) F(t,y) = h(t,y).$$

Hence,

$$p(t,x,y) = -\frac{1}{\gamma}\ln\left(\frac{h(t,y)}{F(t,y)}\right),$$

(9.38)

with $F$ and $h$ given in (9.37) and (9.36), respectively. This price can be computed by Monte Carlo simulations of the paths of the Markov process $Y(t)$. An analogous expression can be derived for the buyer's indifference price.

In [Carmona and Diko (2005)] the claims of the form $\xi = \max(\int_{\tau_1}^{\tau_2} \phi(Y(s))\,ds - K, 0)$ for some strike $K > 0$ are discussed. Such claims, as argued by the authors, can be priced by introducing a new state variable $Z(t) = \int_{\tau_1}^t Y(s)\,ds$, the amount of rainfall from $\tau_1$ up to time $t \geq \tau_1$. Then, the claim is represented as

$$\xi = \int_{\tau_1}^{\tau_2} \mathbf{1}(Z(s) > K)\phi(Y(s))\,ds.$$

Using a function $\hat{g}(t,y,z)$ rather than $g(t,y)$ above will give the desired results.

As an empirical application, [Carmona and Diko (2005)] consider the pricing of call options on the cumulative amount of rainfall in Bergen, Norway. They price these options using electricity forward contracts traded on NordPool, the Nordic power exchange. It is argued empirically that there is a relationship between rainfall intensity in Bergen and the electricity forward contract delivering in the fourth quarter. Using a constant volatility $\sigma(y) = \sigma$ and a non-linear drift $\mu(y) = a\ln(\epsilon + y) + b$ with constants $a, b$ and $\epsilon$, where $\epsilon > 0$ is a cutoff level for no rainfall, [Carmona and Diko (2005)] estimate the parameters in the rainfall model $Y(t)$ and the fourth quarter forward price dynamics $S(t)$ based on data from 2002. We refer the reader to [Carmona and Diko (2005)] for a detailed presentation of the estimation procedure and a discussion of the pricing based on the indifference approach of a call option on cumulative rainfall.

## 9.2     Fair pricing by benchmarking to a reference index

In [Platen and West (2005)] a fair pricing approach is developed for weather derivatives based on benchmarking against a reference portfolio. The method is developed for pricing of derivatives in general incomplete markets, that is, pricing of derivatives in markets where one cannot hedge perfectly the derivative. The idea is to use some portfolio as numéraire, and price the derivatives using a martingale assumption on the discounted prices. We now explain the fair pricing approach in more detail in the context of weather derivatives.

Suppose we have a derivative written on some weather index paying the amount $Y$ at time $\tau > 0$, where $Y$ is some $\mathcal{F}_\tau$-measurable random variable. In the arbitrage pricing theory of mathematical finance (see e.g. [Duffie (1992)]), we know that the price dynamics of the derivative, denoted by $C(t)$, is a martingale under a pricing measure $Q$ after discounting by the risk-free interest rate. In other words, if we use a bank account as numéraire, the pricing $C(t)$ with respect to this, $C(t)/\exp(rt)$ is a $Q$-martingale. Of course, the fundamental condition for stating this is that the derivative $Y$ can be perfectly replicated by a portfolio consisting of the underlying asset and the bank account. This is the complete market situation, which does not cover the case of weather markets since the underlying of $Y$ is trivially not an asset.

Rather than using the bank account as numéraire, [Platen and West (2005)] introduce a reference index being a portfolio optimally invested in the markets. Suppose that the portfolio has the value described by a stochastic process $V^{(\pi)}(t)$, with an investment strategy $\pi$. The *fair price* of $Y$ is defined as the process $C(t)$ such that $C(t)/V^{(\pi)}(t)$ is a martingale. Hence, since $C(\tau) = Y$, we immediately obtain the fair price

$$C(t) = \mathbb{E}\left[ Y \frac{V^{(\pi)}(t)}{V^{(\pi)}(\tau)} \,\middle|\, \mathcal{F}_t \right]. \tag{9.39}$$

Note that the martingale condition is with respect to the market probability $P$, and not any pricing measure $Q$ equivalent to $P$. In fact, [Platen and West (2005)] make a point out of this since their pricing approach does not require the existence of any such pricing measures, and the problem of specifying them is avoided.

The benchmark $V^{(\pi)}(t)$ is chosen to be the *growth optimal portfolio*. We now explain how this portfolio is constructed. Assume a market consisting of $d + 1$ assets with corresponding price dynamics $S_0(t), S_1(t), \ldots, S_d(t)$, where $S_0(t)$ is a risk-free bond (denoted by $D$ in the previous section). Next, let $\pi_i(t)$, $i = 0, \ldots, d$, be the *number* of shares invested in asset $i$, and we assume that the strategy $\pi(t) = (\pi_0(t), \ldots, \pi_d(t))$ gives a self-financing portfolio

$$dV^{(\pi)}(t) = \sum_{i=0}^{d} \pi_i(t) \, dS_i(t) .$$

We implicitly impose conditions on the stochastic integrals $\pi_i(t) \, dS_i(t)$ to make sense. To avoid many technical details, we refrain from presenting these conditions here, but refer the interested reader to [Platen and West (2005)]. To get the growth optimal portfolio, we solve the stochastic control problem

$$\sup_{\pi} \mathbb{E} \left[ \ln \left( V^{(\pi)}(\tau) \right) \mid \mathcal{F}_t \right] ,$$

optimizing terminal wealth under logarithmic utility.

As argued by [Platen and West (2005)], the growth optimal portfolio outperforms most other strategies in the sense of return, and it has a close resemblance with the MSCI World Index[2] when we consider fully diversified portfolios on the financial market. Hence, one can use the MSCI World Index as a proxy for the growth optimal portfolio.

In the paper of [Platen and West (2005)] it is shown statistically that temperature indices in Australia are not correlated to the MSCI World Index. In other words, temperature variations in one location do not influence the world's financial markets, which is not very surprising. On the other hand, this gives a rather simple formula for the fair price of a weather derivative. Since $Y$ is naturally independent of $V^{(\pi)}(t)$, we find that

$$C(t) = \mathbb{E} \left[ \frac{V^{(\pi)}(t)}{V^{(\pi)}(\tau)} \, \bigg| \, \mathcal{F}_t \right] \mathbb{E} \left[ Y \mid \mathcal{F}_t \right] .$$

But in the fair pricing approach of [Platen and West (2005)], the first expectation in this expression is actually the price of a zero-coupon bond at time $t$ and maturity at time $\tau \geq t$, denoted $P(t, \tau)$. Thus, we obtain

---

[2]The MSCI World Index is composed of over 1600 stocks from markets in the developed world.

that the fair price of the weather derivative is given by its so-called actuarial price

$$C(t) = P(t, \tau)\mathbb{E}\left[Y \mid \mathcal{F}_t\right] . \tag{9.40}$$

Note that it is possible to use a stochastic model for the zero-coupon bond prices $P(t, \tau)$. In the case of a constant interest rate, $P(t, \tau) = \exp(-r(\tau - t))$, we are back to the risk-neutral pricing framework used in Chapter 7 with the special choice $Q = P$.

The fair pricing approach provides a rational explanation for why one should price weather derivatives by expected values without any risk adjustments. [Platen and West (2005)] argue that there should not be any risk premium in the market since investors may fully diversify weather risk in space and time. One can diversify using weather derivatives written on different locations in space and at different times. According to [Platen and West (2005)], the premia actually observed in markets today can be explained by lack of liquidity, which they predict will disappear with increasing market activity.

## 9.3    Pricing by marginal utility

Pricing by *marginal utility* was proposed for weather derivatives by [Davis (2001)], and is closely linked to the theory of utility indifference approach presented in Sect. 9.1. As it turns out, it produces pricing formulas for weather derivatives which are similar to the fair prices of [Platen and West (2005)]. We introduce the marginal utility approach following closely the presentations in [Davis (1997, 2001)].

Recall the market in Sect. 9.1 with a risky asset and a bank account, and an investor with wealth $X^\pi(t)$ at time $t$, where the initial wealth $x$ is allocated into the two assets using a strategy $\pi$. Suppose the investor has a utility function $u : \mathbb{R}_+ \mapsto \mathbb{R}$ which is non-decreasing and twice continuously differentiable, $u'(x) > 0$, $\lim_{x \to 0} u'(x) = \infty$ and $\lim_{x \to \infty} u'(x) = 0$. Given an initial wealth $X^\pi(0) = x$, the investor maximizes the expected utility by solving the stochastic control problem

$$U(t, x) = \sup_\pi \mathbb{E}\left[u(X^\pi(\tau)) \mid X^\pi(t) = x\right] , \tag{9.41}$$

where the supremum is taken over all admissible controls. Recall from the utility indifference case (see Sect. 9.1) a weather derivative written on an index with dynamics $Y(t)$, paying $f(Y(\tau))$ at time $\tau$. We derive a price of it by using the argument of *marginal rate of substitution*.

Instead of investing only in the assets available, the investor could place some of the initial endowment in the weather derivative. Let us say the investor allocates $\delta$ amount of the initial wealth to the weather derivative, with the price $p$. The wealth can then be optimized by solving the stochastic control problem

$$V(\delta, t, x, y, p) = \sup_{\pi} \mathbb{E}\left[u\left(X^{\pi}(\tau) + \frac{\delta}{p}f(Y(\tau))\right)\Big| X^{\pi}(t) = x - \delta, Y(t) = y\right].$$
(9.42)

The fair price $p$ is defined as the price on the derivative which makes the marginal maximum expected utility at $\delta = 0$ equal to zero. In other words, the price $p$ such that any changes in $\delta$ at zero have no effect on the value of $V$. In formal terms, we define this as $p = p(t, x, y)$ solving the equation

$$\frac{\partial V}{\partial \delta}(0, t, x, y, p(x, y)) = 0.$$
(9.43)

In [Davis (1997)], it is shown that if $U(t, x)$ is differentiable for $x > 0$ and $U'(t, x) > 0$, it holds that

$$p(t, x, y) = \frac{\mathbb{E}\left[u'(X^{\pi^*}(\tau))f(Y(\tau)) \,|\, X^{\pi^*}(t) = x, Y(t) = y\right]}{U'(t, x)},$$
(9.44)

where, $\pi^*$ is the optimal investment strategy.

We assume that $u(x) = \ln(x)$, a utility function satisfying the conditions above. Let us first derive $U(t, x)$ based on such a utility function. Similar to the case of utility indifference pricing, we get the nonlinear partial differential equation

$$\partial_t U + rx\partial_x U - \frac{(\mu - r)^2}{2\sigma^2}\frac{(\partial_x U)^2}{\partial_{xx} U} = 0,$$

with terminal condition $U(\tau, x) = \ln(x)$. Guessing on a solution of the form $U(t, x) = h(t) + \ln x$, we find that

$$U(t, x) = \left(r + \frac{(\mu - r)^2}{2\sigma^2}\right)(\tau - t) + \ln x.$$
(9.45)

Hence, $U'(t, x) = 1/x > 0$ for all $x > 0$. The fair marginal price then becomes

$$p(t, x, y) = x\mathbb{E}\left[(X^{\pi^*}(\tau))^{-1}f(Y(\tau)) \,|\, X^{\pi^*}(t) = x, Y(0) = y\right].$$
(9.46)

The price may be stated in a more suggestive form as

$$p(t, x, y) = \mathbb{E}\left[\frac{X^{\pi^*}(t)}{X^{\pi^*}(\tau)}f(Y(\tau))\Big| X^{\pi^*}(t) = x, Y(t) = y\right].$$

Notice the similarity with the fair price in the benchmark approach in Sect. 9.2, when using the risky asset correlated with the temperature index as the asset in the growth optimal portfolio. Following the approach of [Davis (1997)], we could use $X^\pi$ as the wealth derived from a portfolio invested in $d$ risky assets and a bank account, and considered a weather derivative with payoff $Z$ rather than $f(Y(\tau))$. Then the same price as for the benchmark approach would be achieved. Hence, one may explain the benchmark approach as coming from marginal rate of substitution with the conclusion that the two pricing approaches coincide.

Notice that the price given by marginal value results in a linear pricing rule. If we buy or sell $n$ derivatives, then the price will be $np(t,x,y)$ of the total derivative purchase. This is in contrast to the pricing rule implied by the indifference approach, which was a highly nonlinear function of the number of derivatives in the trade.

In [Davis (2001)], the particular case of a gas producer with risk exposure to temperature is considered. The producer sells gas per unit time with volume being proportional to the current temperature index $Y(t)$, i.e.,

$$v(t) = cY(t).$$

Here, $Y(t)$ is assumed to follow the dynamics in (9.3). The profit of the producer is $Z(t) = cY(t)S(t)$, where $S(t)$ is the spot price of gas with dynamics given by (9.1). In the setup of [Davis (2001)], the producer does not face any investment decisions, but simply produces up to the current level of demand and sells at market price. Hence, $X^{\pi^*}(t) := Z(t)$, and the price of the temperature derivative becomes

$$p(t,y) = \mathbb{E}\left[\frac{Y(t)S(t)}{Y(\tau)S(\tau)} f(Y(\tau)) \,\Big|\, Y(t) = y\right]. \tag{9.47}$$

We observe that the price is not depending on the proportionality constant $c$ in the linear relation between volume and temperature. In fact, it is not depending on the current gas price $s$ either since the ratio $S(t)/S(\tau)$ is independent of $s$ due to the geometric Brownian motion structure of $S$. Hence, the price $p$ is only a function of current time and level of temperature index $y$.

Let us discuss the pricing rule in (9.47) further. Put $X(t) = S(t)Y(t)$ and observe from an application of Itô's Formula in Thm. 4.1 that

$$dX(t) = (\mu + \alpha + \rho\sigma\beta)X(t)\,dt + X(t)(\sigma dB(t) + \beta dW(t)).$$

Define the Brownian motion

$$dB_0(t) = \frac{1}{\sqrt{\sigma^2 + \beta^2 + 2\rho\sigma\beta}}(\sigma dB(t) + \beta dW(t)), \tag{9.48}$$

together with the constants

$$\mu_0 = \mu + \alpha + \rho\sigma\beta$$

and

$$\sigma_0 = \sqrt{\sigma^2 + \beta^2 + 2\rho\sigma\beta}\,.$$

Then $X(t)$ will have the dynamics

$$dX(t) = \mu_0 X(t)\, dt + \sigma_0\, dB_0(t)\,. \tag{9.49}$$

Observe that,

$$\mathbb{E}[W(t)B_0(t)] = \frac{1}{\sqrt{\sigma^2 + \beta^2 + 2\rho\sigma\beta}}(\sigma\mathbb{E}[W(t)B(t)] + \beta\mathbb{E}[W^2(t)]) = \rho_0 t$$

for

$$\rho_0 = \frac{\rho\sigma + \beta}{\sqrt{\sigma^2 + \beta^2 + 2\rho\sigma\beta}}\,. \tag{9.50}$$

Hence, the two Brownian motions $W$ and $B_0$ are correlated by the factor $\rho_0$. We find that

$$\frac{X(t)}{X(\tau)} = \exp\left(-\sigma_0(B_0(\tau) - B_0(t)) + \frac{1}{2}\sigma_0^2(\tau - t)\right)\exp(-\mu_0(\tau - t))$$

$$= \exp\left(-\sigma_0(B_0(\tau) - B_0(t)) - \frac{1}{2}\sigma_0^2(\tau - t)\right)\exp(-(\mu_0 - \sigma_0^2)(\tau - t))\,.$$

We next apply Girsanov's Theorem (see [Øksendal (1998)]) to introduce a probability $P_0$ equivalent to $P$, with Radon-Nikodym density

$$\left.\frac{dP_0}{dP}\right|_{\mathcal{F}_t} = \exp\left(-\sigma_0 B_0(t) - \frac{1}{2}\sigma_0^2 t\right)\,. \tag{9.51}$$

Since $B_0$ can be represented as

$$B_0(t) = \rho_0 W(t) + \sqrt{1 - \rho^2}\epsilon(t)$$

in distribution, with a Brownian motion $\epsilon(t)$ independent of $W(t)$, we find that

$$dW_0(t) = \rho_0\sigma_0\, dt + dW(t)\,, \tag{9.52}$$

is a Brownian motion under $P_0$. From Bayes' rule in [Karatzas and Shreve (1991)], we find that the price in (9.47) becomes

$$p(t, y) = e^{-(\mu_0 - \sigma_0^2)(\tau - t)}\mathbb{E}_{P_0}\left[f(Y(\tau)) \,|\, Y(t) = y\right]\,, \tag{9.53}$$

where $Y(t)$ has $P_0$-dynamics

$$dY(t) = (\alpha - \rho_0\sigma_0\beta)Y(t)\, dt + \beta Y(t)\, dW_0(t)\,. \tag{9.54}$$

With this at hand, we can express the price in terms of the standard normal distribution function, which is done in the next Proposition.

**Proposition 9.4.** *The marginal price of a weather derivative with payoff $f(Y(\tau))$ at time $\tau$, written on an index with dynamics $Y(t)$ defined in (9.3), is*

$$p(t, y) = e^{-(\mu_0 - \sigma_0^2)(\tau - t)} \int_{\mathbb{R}} f\left(y e^{(\alpha - \rho_0 \sigma_0 - \frac{1}{2}\beta^2)(\tau - t) + \beta\sqrt{\tau - t}z}\right) \phi(z)\, dz\,.$$

*Here, $\phi$ is the standard normal distribution function.*

**Proof.**     From (9.54) we have that under $P_0$,

$$Y(\tau) = y \exp\left(\left(\alpha - \rho_0\sigma_0 - \frac{1}{2}\beta^2\right)(\tau - t) + \beta(W_0(\tau) - W_0(t))\right)\,.$$

Since, in distribution, $W_0(\tau) - W_0(t) = \epsilon\sqrt{\tau - t}$, where $\epsilon$ is a standard normal random variable, the result follows.     $\square$

Let us consider the forward price $F(t, \tau)$ at time $t$ of a contract written on the index $Y$ maturing at time $\tau \geq t$. Since the payoff from a long position in the forward is $Y(\tau) - F(t, \tau)$ with $F(t, \tau)$ naturally $\mathcal{F}_t$-adapted, we find from (9.53) that

$$F(t, \tau) = \mathbb{E}_{P_0}[Y(\tau) \,|\, Y(t)]\,.$$

After a direct computation, we find the forward price to be

$$F(t, \tau) = Y(t)e^{(\alpha - \beta - \rho\sigma)(\tau - t)}\,. \tag{9.55}$$

Hence, the marginal utility pricing approach provides us with a dynamics for the forward price which explicitly depends on today's value of the underlying index $Y(t)$ discounted by a "convenience yield" factor $\alpha - \beta - \rho\sigma$.

Applying Itô's Formula in Thm. 4.1 and using the $P$-dynamics of $Y$ in (9.3), we find that

$$dF(t, \tau) = (\beta + \rho\sigma)F(t, \tau)\, dt + \beta F(t, \tau)\, dW(t)\,.$$

Introduce the process

$$d\widetilde{W}(t) = \left(1 + \rho\frac{\sigma}{\beta}\right) dt + dW(t)\,, \tag{9.56}$$

which is a Brownian motion under a probability $Q$ equivalent to $P$ by Girsanov's Theorem (see [Øksendal (1998)]). We find that

$$dF(t, \tau) = \beta F(t, \tau)\, d\widetilde{W}(t)\,,$$

and therefore $Q$ is the risk-neutral probability for the forward price. Hence, the marginal utility pricing approach produces a *market price of risk*

$$\theta = 1 + \rho\frac{\sigma}{\beta}. \tag{9.57}$$

If we think of the index dynamics $Y(t)$ in (9.3) as an approximation of the true dynamics coming from a temperature model as we have discussed in Chapter 4, then the marginal utility pricing approach provides us with an explicitly given market price of risk. The market price of risk $\theta$ is positive whenever the tradeable asset price $S$ is positively correlated with the index $Y$. If $\rho < 0$, on the other hand, then we may get a negative market price of risk as long as $\rho\sigma < -\beta$. In Chapter 5 forward prices on temperature and wind speed under pricing measures $Q$ parametrized by the market prices of risk $\theta$ were derived.

We remark that under the risk-neutral dynamics for the forward price, we can find a formula for the price of European call options. Since the forward price $F(t,\tau)$ follows a geometric Brownian motion dynamics with constant volatility $\beta$, the price can be derived using the Black-76 formula as in Prop. 7.6.

One may wonder to what extent a weather index can be modelled as a geometric Brownian motion, as it is assumed in the approaches discussed above. Let us focus on the three most used temperature indices at the CME market, namely HDD, CDD and CAT. The CAT index is an aggregation of temperatures over an interval. The empirical study of Lithuanian data in Chapter 3 strongly points towards normality of temperatures $T(t)$ at each time instance $t$, and therefore the CAT index over a measurement period $[\tau_1, \tau_2]$

$$\int_{\tau_1}^{\tau_2} T(s)\,ds$$

is normally distributed as well. In the above context, it is natural to assume that

$$Y(t) = \int_{t-(\tau_2-\tau_1)}^{t} T(s)\,ds$$

follows a geometric Brownian motion dynamics. Note in passing that $Y(\tau_2)$ coincides with the CAT index over $[\tau_1, \tau_2]$ and that $Y$ is the moving CAT index over the intervals $[t-(\tau_2-\tau_1), t]$. Moreover, after a change of variables, we can represent $Y(t)$ alternatively as

$$Y(t) = \int_0^{\tau_2-\tau_1} T(t-s)\,ds.$$

As the model for $Y(t)$ assumes lognormally distributed values, a big deviation between the actual CAT index and $Y(t)$ for any time instance $t$ can appear.

Concerning the HDD and CDD indices, we are at least sure that they have positive values, which are skewed. However, both indices will have a positive mass at zero, coinciding with the probability that temperature is less than the threshold value for all times over the measurement period. The geometric Brownian motion will not have any such mass at zero. In fact, the probability of $Y(t) = 0$ is zero at any time instance $t$. The longer the measurement period, the more unlikely it is that the temperature is always below the threshold. This, of course, also depends on the temperature pattern at a given location and the value of the threshold itself.

Let us make a simulation-based study of the CDD index, using a simple Ornstein-Uhlenbeck model for the deseasonalized temperature $T(t) - \Lambda(t) = X_1(t)$, with

$$dX_1(t) = -\alpha X_1(t)\, dt + \sigma dB(t)\,. \tag{9.58}$$

Assume further that the seasonality function is constant and equal to 20, i.e., $\Lambda(t) = 20$. With the usual threshold of $c = 18°\text{C}$ for the CDD index, the temperature dynamics mean-reverts at around a level of $2°\text{C}$ above the threshold. A constant seasonality level is not true for most locations and is used here only for simplicity. We define the CDD index at time $t$ over a measurement period of $\tau_2 - \tau_1$ days, $\tau_2 > \tau_1$, as

$$\text{CDD}(t) := \sum_{i=0}^{\tau_2 - \tau_1} \max(X_1(t - i) + 2, 0)\,. \tag{9.59}$$

The aim is to check the validity of a geometric Brownian motion approximation to the dynamics of $\text{CDD}(t)$.

In order to simulate the CDD index, we simulate sample paths of $X_1(t)$ at discrete times $t$. For that purpose, we use the explicit dynamics of $X_1$ to get

$$X_1(t + 1) = X_1(t)\exp(-\alpha) + \int_t^{t+1} \sigma \exp(-\alpha(t + 1 - s))\, dB(s)\,.$$

Observe that the stochastic integral is independent of $X_1(t)$, and normally distributed with mean zero and variance $\sigma^2(1 - \exp(-2\alpha))/2\alpha$. In the empirical study, we choose $\sigma = 2$ and $\alpha = 0.2$, values reasonable for a temperature dynamics. In Fig. 9.1 we plot a sample path of $\text{CDD}(t)$ for 1000 days, using a 30 days measurement period. We initialized the CDD index

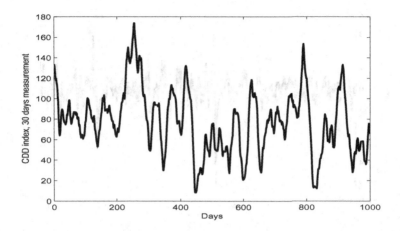

Fig. 9.1    Simulated path of the CDD with 30 days measurement period.

by first simulating 30 days of temperatures, and next iteratively simulated the path. The initial temperature was set to the mean level, $T(0) = 20°C$.

Since a geometric Brownian motion has log returns which are independent and identically normally distributed, we investigate the log returns from the sample path of the CDD. In Fig. 9.2 we observe the correponding log returns, calculated as

$$\ln\left(\mathrm{CDD}(t+1)\right) - \ln\left(\mathrm{CDD}(t)\right)$$

with time running through the days $t = 1, \dots, 999$. The empirical density function of the log return data is plotted in Fig. 9.3 together with the best fit of a normal distribution (dotted line). The empirical density function was computed using the built-in function `ksdensity` in Matlab. The mean and standard deviation of the normal fit were estimated to be $-0.000735$ and $0.0633$, respectively. As it is seen from the densities, the normal distribution assigns too little probability to the small log returns, and too much for "intermediate sized" log returns. The tails of the normal distribution are much lighter than those coming from the simulated one. For example, the empirical 1% and 99% quantiles of the log return data are $-0.174$ and $0.160$, respectively, while the same quantiles for the normal distribution are $-0.148$ and $0.147$. Finally, in the ACF of the log returns in Fig. 9.4, there are clear signs of dependency in the log return data. For small lags, correlations between log return data are high, which we attribute to the fact that the index dynamics is defined as a rolling window. In addition, the

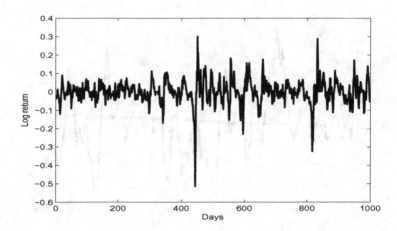

Fig. 9.2     Log returns from the simulated path of CDD with 30 days measurement period.

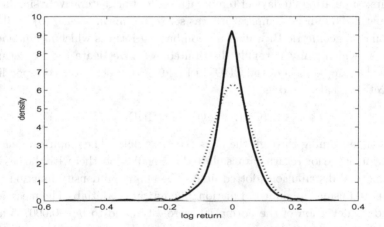

Fig. 9.3     Density plot of the log returns from the simulated path of CDD (smooth line) together with the fitted normal density (dotted line).

mean reversion effect in the temperature dynamics adds to this memory. Note that since we have specified the seasonality function to be above the threshold, we do not get any zero values for the CDD index. However, if we consider a seasonality equal to the threshold, the simulation study shows occurrences of a CDD index with zero value. In this case the log returns will not be defined, which clearly does not coincide with a geometric Brownian

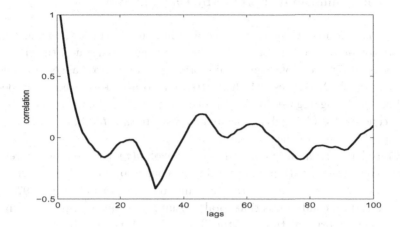

Fig. 9.4    ACF for the log returns from the simulated path of CDD.

motion hypothesis.

Remark that we have made several simplifying assumptions in our simulation study. First and foremost, the seasonality function of the temperature is at most locations varying with time, not a constant. Secondly, as it was demonstrated in Chapter 3, the temperature dynamics is following a higher-order AR process, with a seasonal volatility. These properties of the temperature dynamics would imply seasonality and additional memory effects in the dynamics of the CDD index, further violating the geometric Brownian motion hypothesis. Despite the arguments against modelling a temperature index using a geometric Brownian motion, the process may still make sense as a first-order approximation. It has the clear advantage of providing simple pricing rules for forwards and options on weather contracts, a very attractive feature in practical situations.

As a final note, we remark that [Stoll and Wiebauer (2010)] detected a strong relationship between gas prices and temperature. Based on CAR models for temperature, the authors identified the seasonal level of the gas spot price as a function of the HDD index. They calibrated the model to gas price data and applied the model to gas storage valuation and pricing of supply contracts. In our context, this study points towards a strong relationship between gas prices and temperatures. One may suspect that there are similar connections between temperature and electricity, and maybe even wind as this becomes gradually more important as a renewable source of energy.

## 9.4    The equilibrium approach by Cao and Wei

In this last Section, we present the equilibrium approach by [Cao and Wei (2004)] for pricing weather derivatives which is an extension of the theory of [Lucas (1978)] for contingent claim pricing in a pure exchange economy. [Cao and Wei (2004)] restrict their attention to a two-state economy described by the aggregate dividend process $\delta(t)$ and the temperature $T(t)$. The economy is evolving dynamically in a discrete time $t = 0, 1, \ldots, \tau$, with the time horizon $\tau$ of the economy.

Consider a claim on temperature with payoff $q(\tau)$ at time $\tau$, and a representative investor with period-$\tau$ utility on consumption $U(c(\tau), \tau)$, where $c(t)$ is the consumption at time $t$. According to the theory of [Lucas (1978)], the equilibrium condition is that total consumption equals aggregated dividend, and the price at time $t$ of the claim, $X(t, \tau)$, is given by

$$X(t,\tau) = \frac{1}{U_c(\delta(t),t)} \mathbb{E}\left[U_c(\delta(\tau),\tau)q(\tau) \,|\, \mathcal{F}_t\right], \qquad (9.60)$$

with $U_c$ being the derivative of $U$ with respect to consumption.

Consider now a temperature futures written on the CDD index over the measurement period $[\tau_1, \tau_2]$, denoted $\text{CDD}(\tau_1, \tau_2)$. Since the price to enter a futures contract is zero, and the payoff from a long position at time $\tau_2$ will be $\text{CDD}(\tau_1, \tau_2) - F_{\text{CDD}}(t, \tau_1, \tau_2)$, we find

$$\mathbb{E}\left[U_c(\delta(\tau_2),\tau_2)(\text{CDD}(\tau_1,\tau_2) - F_{\text{CDD}}(t,\tau_1,\tau_2)) \,|\, \mathcal{F}_t\right] = 0.$$

Naturally, the futures price $F_{\text{CDD}}(t, \tau_1, \tau_2)$ is $\mathcal{F}_t$-adapted, since at time $t$ it is only possible to determine this price based on the information $\mathcal{F}_t$. Hence,

$$F_{\text{CDD}}(t,\tau_1,\tau_2) = \mathbb{E}\left[\frac{U_c(\delta(\tau_2),\tau_2)}{\mathbb{E}[U_c(\delta(\tau_2),\tau_2) \,|\, \mathcal{F}_t]}\text{CDD}(\tau_1,\tau_2) \,\bigg|\, \mathcal{F}_t\right]. \qquad (9.61)$$

In order to compute the futures price, one must specify a model for the temperature $T$, the aggregated dividend $\delta$ and the utility $U$. [Cao and Wei (2004)] assume a power utility on consumption defined by

$$U(c,t) = e^{-\rho t}\frac{c^{\gamma+1}}{\gamma+1}, \qquad (9.62)$$

where $\gamma \in (-\infty, 0]$ is the risk parameter and $\rho > 0$ is the rate of time preference. With this specification,

$$U_c(c,t) = e^{-\rho t}c^\gamma,$$

and therefore the futures price becomes

$$F_{\text{CDD}}(t,\tau_1,\tau_2) = \mathbb{E}\left[\frac{\delta^\gamma(\tau_2)}{\mathbb{E}[\delta^\gamma(\tau_2) \,\big|\, \mathcal{F}_t]}\text{CDD}(\tau_1,\tau_2) \,|\, \mathcal{F}_t\right].$$

As a model for the aggregate dividend, [Cao and Wei (2004)] use an exponential AR(1) process with an ARCH structure for the noise.

Recall the marginal and benchmark approaches discussed in the Sections above. As we see, the price of a temperature futures based on equilibrium arguments has the same structure as for these two approaches, except that the ingredients in the pricing formulas have different intepretations. Noteworthy is that all three approaches suggest a pricing measure $Q$ depending on external effects, and not restricted to a change of measure which only involves the temperature dynamics itself. In our analysis, we restrict attention to pricing measures $Q$ defined in terms of the temperature dynamics alone, in this way not bringing in any external sources of uncertainty. In a Brownian setup in Chapters 4 and 5, one could easily extend the measure change to also include dependency on external uncertainties described by stochastic processes. The possibilities are only restricted by the (rather general) conditions of the Girsanov Theorem (see [Øksendal (1998)]).

The theoretical framework in [Cao and Wei (2004)] is applied to price various temperature derivatives. Their discrete-time model for temperature shares many similarities with the CARMA models suggested in Sect. 4.6. Moreover, they estimate the model on temperature data from US cities, and apply the Gross national product of the US as a proxy for the aggregated dividend. The weather risk premium is discussed for futures and options. We refer to [Cao and Wei (2004)] for the details on the analysis.

# Appendix A

# List of abbreviations

ACF – *autocorrelation function*
AIC – *Akaike's Information Criteria*
AR – *autoregressive*
ARCH – *autoregressive conditional heteroskedastic*
ARFIMA – *autoregressive fractionally integrated moving average*
ARMA – *autoregressive moving average*
BNS – *Barndorff-Nielsen and Shephard*
BSS – *Brownian semistationary*
°C – *degrees Celsius*
CAR – *continuous autoregressive*
CARMA – *continuous autoregressive moving average*
CAT – *cumulative average temperature*
CDD – *cooling-degree day*
CHI – *CME Hurricane Index*
CME – *Chicago Mercantile Exchange*
EEX – *European Electricity Exchange*
EUR – *Euro*
FFT – *fast Fourier transform*
GBP – *British Pounds*
GARCH – *generalized autoregregressive conditional heteroskedastic*
°F – *degrees Fahrenheit*
HDD – *heating-degree days*
HJB – *Hamilton-Jacobi-Bellman*
ICE – *Inter Continental Exchange*
KS – *Kolmogorov-Smirnov*
LIFFE – *London International Financial Future and Option Exchange*
MA – *moving average*

OTC – *over-the-counter, bilateral "market"*

PACF – *partial autocorrelation function*

PRIM – *Pacific Rim*

RCLL – *right continuous with left limits*

QCARGH – *quadratic generalized autoregressive conditional heteroskedastic*

SARMAR – *spatial autoregressive moving average regression*

STAR – *spatial-temporal autoregressive*

STARMA – *spatial-temporal autoregressive moving average*

USD – *US Dollars*

USFE – *US Futures Exchange*

WRMA – *Weather Risk Management Association*

# Bibliography

Abramowitz, M., and Stegun, I. A. (eds.) (1965). *Handbook of Mathematical Functions with Formulas, Graphs, and Mathematical Tables*, New York: Dover.

Ailliot, P., Monbet, V., and Prevosto, M. (2006). An autoregressive model with time-varying coefficients for wind fields, *Environmetrics* **17**, pp. 107–117.

Alaton, P., Djehiche, B., and Stillberger, D. (2002). On modelling and pricing weather derivatives, *Appl. Math. Finance* **9**, pp. 1–20.

Alos, E., Mazet, O., and Nualart, D. (2001). Stochastic calculus with respect to Gaussian processes, *Ann. Probability* **29**(2), pp. 766–801.

Ankirchner, S., Imkeller, P., and Dos Reis, G. (2010). Pricing and hedging of derivatives based on nontradeable underlyings, *Math. Finance* **20**(2), pp. 289–312.

Barndorff-Nielsen, O. E., Benth, F. E., and Veraart, A. E. D. (2010). Modelling energy spot prices by Lévy semistationary processes, *CREATES Research Paper 201018*. Available at SSRN: http://ssrn.com/abstract=1597700.

Barndorff-Nielsen, O. E., Benth, F. E., and Veraart, A. E. D. (2012). Modeling electricity forward markets by ambit fields, *Submitted manuscript*.

Barndorff-Nielsen, O. E., and Schmiegel, J. (2004). Lévy-based tempo-spatial modelling; with applications to turbulence, *Uspekhi Mat. Nauk* **59**, pp. 65–91.

Barndorff-Nielsen, O. E., and Schmiegel, J. (2007). Ambit processes: with applications to turbulence and cancer growth. In Benth, F. E., Di Nunno, G., Lindstrøm, T., Øksendal, B., and Zhang, T. (eds.), *Stochastic Analysis and Applications: The Abel Symposium 2005*, Springer Verlag, Heidelberg, pp. 93–124.

Barndorff-Nielsen, O. E., and Schmiegel, J. (2009). Brownian semistationary processes and volatility/ intermittency. In Albrecher, H., Rungaldier, W., and Schachermeyer, W. (eds.), *Advanced Financial Modelling*, Radon Series on Computational and Applied Mathematics 8, W. de Gruyter, Berlin, pp. 1–26.

Barndorff-Nielsen, O. E., and Shephard, N. (2001). Non-Gaussian Ornstein-Uhlenbeck-based models and some of their uses in economics, *J. R. Statist. Soc. B* **63**(2), pp. 167–241 (with discussion).

Barnett, B. J., Barrett, C. B., and Skees, J. R. (2008). Poverty traps and index based risk transfer products, *World Development* **36**, pp. 1766–1785.

Barnett, B. J., and Mahul, O. (2007). Weather index insurance for agriculture and rural areas in lower income countries, *Amer. J. Agriculture Econ.* **89**, pp. 1241–1247.

Barth, A., Benth, F. E., and Potthoff, J. (2011). Hedging of spatial temperature risk with market-traded futures, *Appl. Math. Finance* **18**(2), pp. 93–117.

Benth, F. E. (2003). On arbitrage-free pricing of weather derivatives based on fractional Brownian motion, *Appl. Math. Finance* **10**, pp. 303–324.

Benth, F. E. (2004). *Option Pricing with Stochastic Analysis: an Introduction to Mathematical Finance*, Springer-Verlag, Berlin Heidelberg.

Benth, F. E. (2010). On forward price modelling in power markets. In Alternative Investments and Strategies, Kiesel, R., Scherer, M., and Zagst, R. (eds.), World Scientific, Chapter 5, pp. 93–122.

Benth, F. E., Härdle, W., and Lopez Cabrera, B. (2011). Pricing of Asian temperature risk. In *Statistical Tools for Finance and Insurance*, Cizek, P., Härdle, W., and Weron, R. (eds.), Springer-Verlag Berlin Heidelberg, Chapter 5, pp. 163–199.

Benth, F. E., Klüppelberg, C., Müller, G., and Vos, L. (2011). Futures pricing in electricity markets based on stable CARMA spot models, *Submitted manuscript*, arXiv:1201.1151.

Benth, F. E., Lange, N., and Myklebust, T. (2012). Pricing energy quanto options, *Manuscript in preparation*.

Benth, F. E., and Šaltytė-Benth, J. (2005). Stochastic modelling of temperature variations with a view towards weather derivatives, *Appl. Math. Finance* **12**(1), pp. 53–85.

Benth, F. E., and Šaltytė-Benth, J. (2007). The volatility of temperature and pricing of weather derivatives, *Quantit. Finance* **7**(5), pp. 553–561.

Benth, F. E., and Šaltytė Benth, J. (2009). Dynamic pricing of wind futures, *Energy Econ.* **31**(1), pp. 16–24.

Benth, F. E., and Šaltytė Benth, J. (2011). Weather derivatives and stochastic modelling of temperature, *Intern. J. Stoch. Anal.* **2011**, Article ID 576791, 21 pages.

Benth, F. E., Šaltytė Benth, J., and Koekebakker, S. (2008). *Stochastic Modelling of Electricity and Related Markets*, World Scientific, Singapore.

Betz, A. (1966). *Introduction to the Theory of Flow Machines*, D. G. Randall, Trans. Oxford, Pergamon Press.

Björk, T. (1998). *Arbitrage Theory in Continuous Time*, Oxford University Press.

Black, F. (1976). The pricing of commodity contracts, *J. Financial Econ.* **3**, pp. 167–179.

Bouette, J.-C., Chassagneux, J.-F., Sibai, D., Terron, R., and Charpentier, A. (2006). Wind in Ireland: long memory or seasonal effect?, *Stoch. Environ. Research Risk Assessment* **20**, pp. 141–151.

Brémaud, P. (1981). *Point Processes and Queues: Martingale Dynamics*, Springer-Verlag, New York.

Brett, A. C., and Tuller, S. E. (1991). The autocorrelation of hourly wind speed observations, *J. Appl. Meteorology* **30**(6), pp. 823–833.

Brockwell, P. J. (2001). Lévy-driven CARMA processes, *Ann. Inst. Statist. Math.* **53**(1), pp. 113–124.

Brockwell, P. J., Davis, R. A., and Yang, Y. (2007). Estimation for non-negative Lévy-driven CARMA processes, *J. Appl. Probab.* **44**(4), pp. 977–989.

Brody, D. C., Syroka, J., and Zervos, M. (2002). Dynamical pricing of weather derivatives, *Quantit. Finance* **3**, pp. 189–198.

Brown, B. G., Katz, R. W., and Murphy, A. H. (1984). Time series models to simulate and forecast wind speed and wind power, *J. Climate Applied Meteorology* **23**(8), pp. 1184–1195.

Brown, P. E., Diggle, P. J., Lord, M. E., and Young, P. C. (2001). Space-time callibration of radar rainfall data, *Appl. Statist.* **50**(2), pp. 221–241.

Bruno, F., Guttorp, P., Sampson, P. D., and Cocchi, D. (2009). A simple non-separable, non-stationary spatiotemporal model for ozone, *Environ. Ecol. Stat.* **16**, pp. 515–529.

Caballero, R., Jewson, S., and Brix, A. (2002). Long memory in surface air temperature: detection, modeling, and application to weather derivative valuation, *Climate Research* **21**, pp. 127–140.

Campbell, S. D., and Diebold, F. X. (2005). Weather forecasting for weather derivatives, *J. Amer. Stat. Assoc.* **100**(469), pp. 6–16.

Cao, M., and Wei, J. (2004). Weather derivatives valuation and market price of risk, *J. Futures Markets* **24**(11), pp. 1065–1089.

Caporin, M., Preś, M., and Torro, H. (2012). Model based Monte Carlo pricing of energy and temperature quanto options. To appear in *Energy Econ.*

Carmona, R., and Diko, P. (2005). Pricing precipitation based derivatives, *Intern. J. Theor. Appl. Finance* **8**(7), pp. 959–988.

Carmona, R. (ed.) (2009). *Indifference Pricing: Theory and Applications*, Princeton Series in Financial Engineering, Princeton University Press: Princeton and Oxford.

Carr, P., and Madan, D. B. (1998). Option valuation using the fast Fourier transform, *J. Comp. Finance* **2**, pp. 61–73.

Carroll, R. J., Chen, R., George, E. I., Li, T. H., Newton, H. J., Schmiediche, H., and Wang, N. (1997). Ozone exposure and population density in Harris County, Texas, *J. Amer. Stat. Assoc.* **92**, 438, pp. 392–404.

Castino, F., Festa, R., and Ratto, C. F. (1998). Stochastic modelling of wind velocities time series, *J. Wind Engin. Indust. Aerodynamics* **74-76**, pp. 141–151.

Celik, A. N. (2003). A statistical analysis of wind power density based on the Weibull and Rayleigh models at the southern region of Turkey, *Renewable Energy* **29**, pp. 593–604.

Chandler, R. E., and Bate, S. (2007). Inference on clustered data using the independence loglikelihood, *Biometrika* **94**(1), pp. 167–183.

Chatfield, C. (2000). *Time-series Forecasting*, Chapman & Hall/CRC.

Clewlow, L., and Strickland, C. (2000). *Energy Derivatives: Pricing and Risk Management*, Lacima Publications, London.

Cliff, A. D., and Ord, J. K. (1975). Space-time modelling with an application to regional forecasting, *Trans. Inst. British Geographers* **64**, pp. 119–128.

Cont, R., and Tankov, P. (2004). *Financial Modelling with Jump Processes*, Chapman & Hall, Boca Raton.

Corotis, R. B., Sigl, A. B., and Klein, J. (1978). Probability models of wind velocity magnitude and persistence, *Solar Energy* **20**, pp. 483–493.

Cowpertwait, P. S. P. (1995). A generalized spatial-temporal model of rainfall based on a clustered point process, *Proc. R. Soc. Lond. A* **450**, pp. 163–175.

Cox, D. R., and Isham, V. (1988). A simple spatial-temporal model of rainfall, *Proc. R. Soc. Lond. A* **415**, pp. 317–328.

Cressie, N. A. C. (1993). *Statistics for Spatial Data*, Wiley & Sons.

Cressie, N. A. C. (1994). Comment on the paper by Handcock and Wallis "An approach to statistical spatial-temporal modeling of meteorlogical fields", *J. Amer. Stat. Assoc.* **90**(431), pp. 379–382.

Cressie, N. A. C., and Huang, H.-C. (1999). Classes of nonseparable, spatio-temporal stationary covariance functions, *J. Amer. Stat. Assoc.* **94**(448), pp. 1330–1340.

Cressie, N. A. C., and Wikle, C. K. (2002). Space-time Kalman filter. In *Encyclopedia of Environmetrics*, El-Shaarawi A. H. and Piegorsch W. W. (eds.), **4**, John Wiley & Sons, Chichester, pp. 2045–2049.

Cripps, E., Nott, D., Dunsmuir, W. T. M., and Wikle C. (2005). Space-time modelling of Sydney harbour winds, *Australian and New Zealand J. Statist.* **47**(1), pp. 3–17.

Dalezios, N. R., and Adamowski, K. (1995). Spatio-temporal precipitation modelling in rural watersheds, *J. Hydrol. Sciences* **40**(5), pp. 553–568.

Da Prato, G., and Zabzcyk, J. (1992). *Stochastic Equations in Infinite Dimensions*, Encyclopedia of Mathematics and Its Applications, 44. Cambridge University Press.

Davis, M. H. A. (1997). Option pricing in incomplete markets. In *Mathematics of Derivative Securities*, Dempster, M. A. H., and Pliska, S. R. (eds.), Cambridge University Press.

Davis, M. H. A. (2001). Pricing weather derivatives by marginal value, *Quantit. Finance* **1**, pp. 305–308.

De Iaco, S., Myers, D. E., and Posa, D. (2002). Nonseparable space-time covariance models: some parametric families, *Math. Geology* **34**(1), pp. 23–42.

Denison, D. G. T., Dellaportas, P., and Mallick, B. K. (2001). Wind speed prediction in a complex terrain, *Environmetrics* **12**, pp. 499–515.

Deutsch, C. V., and Journel, A. G. (1998). *Gslib: Geostatistical Software Library and User's Guide*, Oxford University Press, New York.

Di Nunno, G., Øksendal, B., and Proske, F. (2009), *Malliavin Calculus for Lévy Processes with Applications to Finance*, Springer Verlag.

Dimitrakopoulos, R. E. (1994). *Geostatistics for the Next Century*, International Forum: Selected Papers. Kluwer Academic.

Dimitrakopoulos, R., and Luo, X. (1997). Joint space-time modelling in the presence of trends. In *Geostatistics Wollongong '96, Vol.1*, Baaffi, E., and Schoefield, N. (eds.), Kluwer Academic Publishers, Dordrecht, pp. 138–149.

Doob, J. L. (1944). The elementary Gaussian processes, *Ann. Math. Statist.* **15**(3), pp. 229–282.

Dornier, F., and Querel, M. (2000). Caution to the wind, *Energy & Power Risk Manag.*, Weather risk special report, August, pp. 30–32.

Duffie, D. (1992). *Dynamic Asset Pricing Theory*, Princeton University Press, Princeton.

Ethier, S. N., and Kurtz, T. G. (1986). *Markov Processes – Characterization and Convergence*, John Wiley & Sons.

Fernandez-Aviles, G., Montero, J. -M., Mateu, J. (2011). Mathematical genesis of the spatio-temporal covariance functions, *J. Math. Statistics* **7**(1), pp. 37–44.

Folland, G. B (1984). *Real Analysis*, Wiley, Chichester.

Franses, P. H., Neele, J., and van Dijk, D. (2001). Modelling asymmetric volatility in weekly Dutch temperature data, *Environ. Modelling Software* **16**, pp. 131–137.

Fuentes, M. (2006). Testing for separability of spatial-temporal covariance functions, *J. Statist. Planning Inference* **136**, pp. 447–466.

Fuentes, M., Chen, L., and Davis, J. M. (2005). A class of nonseparable and nonstationary spatial temporal covariance functions, *Environmetrics* **19**(5), pp. 487–507.

Garcia, A., Torres, J. L., Prieto, E., and De Francisco, A. (1998). Fitting wind speed distributions: a case study, *Solar Energy* **62**(2), pp. 139–144.

Garcia, I., Klüppelberg, C., and Müller, G. (2010). Estimation of stable CARMA models with an application to electricity spot prices, *Statist. Modelling* **11**(5), pp. 447–470.

Geman, H. (1999). *Insurance and Weather Derivatives*, RISK Books.

Geman, H., and Leonardi, M. P. (2005). Alternative approaches to weather derivatives valuation, *Manag. Finance* **31**(6), pp. 46–72.

Glasserman, P. (2003). *Monte Carlo Methods in Financial Engineering*, Springer Verlag.

Gneiting, T. (2002). Nonseparable, stationary covariance functions for space-time data, *J. Amer. Stat. Assoc.* **97**(458), pp. 590–600.

Gneiting, T., Genton, M. G., and Guttorp, P. (2007). Geostatistical space-time models, stationarity, separability and full symmetry, In *Statistical Methods for Spatio-Temporal Systems*, Finkenstadt, B., Held, L., and Isham, V. (eds.), Boca Raton: Chapman & Hall/CRC, pp. 151–175.

Guenni, L., and Hutchinson, M. F. (1998). Spatial interpolation of the parameters of a rainfall model from ground-based data, *J. Hydrology* **212-213**, pp. 335–347.

Handcock, M. S., and Wallis, J. R. (1994). An approach to statistical spatial-temporal modeling of meteorological fields, *J. Amer. Stat. Assoc.* **89**(426), pp. 368–378.

Härdle, W., and Lopez Cabrera, B. (2012). The implied market price of weather risk, *Appl. Math. Finance* **19**(1), pp. 59–95.

Härdle, W., Lopez Cabrera, B., Ohkrin, O., and Wang, W. (2010). Localizing temperature risk, *Submitted manuscript*.

Haslett, J., and Raftery, A. E. (1989). Space-time modelling with long-memory dependence: assessing Ireland's wind power resource, *Appl. Statist.* **38**(1), pp. 1–50.

Heath, D., Jarrow, R. and Morton, A. (1992). Bond pricing and the term structure of interest rates: a new methodology for contingent claim valution, *Econometrica* **60**, pp. 77–105.

Høst, G., Omre, H., and Switzer, P. (1995). Spatial interpolation errors in monitoring data, *J. Amer. Stat. Assoc.* **90**(431), pp. 853–861.

Huang, Z., and Chalabi, Z. S. (1995). Use of time-series analysis to model and forecast wind speed, *J. Wind Engin. Indust. Aerodynamics* **56**, pp. 311–322.

Huang, H.-Ch., and Cressie, N. A. C. (1996). Spatio-temporal prediction of snow water equivalent using the Kalman filter, *Comp. Statist. Data Analaysis* **22**, pp. 159–175.

Hull, J. (2002). *Options, Futures and Other Derivatives*, Prentice-Hall.

Ikeda, N., and Watanabe, S. (1981). *Stochastic Differential Equations and Diffusion Processes*, North-Holland/Kodansha.

Im, H.-K., Rathouz, P. J., and Frederick, J. E. (2009). Space-time modeling of 20 years of daily air temperature in the Chicago metropolitan region, *Environmetrics*, **20**, pp. 494–511.

IPCC AR4 WG1 (2007). Climate Change 2007: The Physical Science Basis, Contribution of Working Group I to the Fourth Assessment Report of the Intergovernmental Panel on Climate Change, Solomon, S., Qin, D., Manning, M., Chen, Z., Marquis, M., Averyt, K. B., Tignor, M., and Miller, H. L. (eds.), Cambridge University Press, ISBN 978-0-521-88009-1, http://www.ipcc.ch/publications_and_data/ar4/wg1/en/contents.html (pbk: 978-0-521-70596-7).

Jacod, J., and Shiryaev, A. N. (1987). *Limit Theorems for Stochastic Processes*, Springer Verlag.

Jewson, S., and Brix, A. (2005). *Weather Derivative Valuation: The Meteorological, Statistrical, Financial and Mathematical Foundations*, Cambridge University Press, Cambridge.

Karatzas, I., Ocone, D. L., and Li, J. (1991). An extension of Clark's formula, *Stoch. Stoch. Rep.* **37**(3), pp. 215–258.

Karatzas, I., and Shreve, S E. (1991). *Brownian Motion and Stochastic Calculus*, Second Edition, Springer Verlag, New York.

Kloeden, P. E., and Platen, E. (1992). *Numerical Solution of Stochastic Differential Equations*, Springer Verlag, Berlin Heidelberg.

Kyriakidis, P. C., and Journel, A. G. (1999). Geostatistical space-time models: a review, *Math. Geology* **31**(6), pp. 651–684.

Leobacher, G., and Ngare, P. (2011). On modelling and pricing rainfall derivatives with seasonality, *Appl. Math. Finance* **18**(1), pp. 71–91.

Li, K. H., Le, N. D., Sun, L., and Zidek, J. V. (1999). Spatial-temporal models for ambient hourly $PM_{10}$ in Vancouver, *Environmetrics* **10**, pp. 321–338.

Loader, C., and Switzer, P. (1989). Spatial covariance estimation for monitoring data, *Study on Statistics and Environmental Factors in Health*, Stanford University. Technical report No. **133**. Downlowded from http://statistics.stanford.edu/~ckirby/techreports/SIMS/, accessed 20/09/2010.

Loader, C., and Switzer, P. (1992). Spatial covariance estimation for monitoring data, In *Statistics in Environmental and Earth Sciences*, Walden, A., and Guttorp, P. (eds.), London: Edward Arnold, pp. 52–70.

Lucas, R. E. (1978). Asset prices in an exchange economy, *Econometrics* **46**, pp. 1429–1445.

Lun, I. Y. F., and Lam, J. C. (2000). A study of Weibull parameters using long-term wind observations, *Renewable Energy* **20**, pp. 145–153.

Ma, Ch. (2003). Families of spatio-temporal stationary covariance models, *J. Statistical Planning and Inference* **116**, pp. 489–501.

Ma, Ch. (2003). Nonstationary covariance functions that model space-time interactions, *Statistics & Probability Letters* **61**, pp. 411–419.

Malliavin, P., and Thalmaier, A. (2006). *Stochastic Calculus of Variations in Mathematical Finance*, Springer-Verlag.

Mardia, K. V., and Goodall, C. R. (1993). Spatial-temporal analysis of multivariate environmental monitoring data, In *Multivariate Environmental Statistics*, Patil, G. P., and Rao, C. R. (eds.), Elsevier Science Publishing, pp. 347–386.

Martin, M., Cremades, L. V., and Santabarbara, J. M. (1999). Analysis and modelling of time series of surface wind speed and direction, *Intern. J. Climatology* **19**, pp. 197–209.

Mimkou, M. (1983). Daily precipitation occurrences modelling with Markov chain of seasonal order, *J. Hydrol. Sciences* **28**(2), pp. 221–232.

Musiela, M., and Zariphopoulou, T. (2004). An example of indifference prices under exponential preferences, *Finance Stoch.* **8**(2), pp. 229–239.

Myers, D. E., and Journel, A. (1990). Variograms with zonal anisotropies and noninvertible kriging systems, *Math. Geology* **22**(7), pp. 779–785.

Niu, X. and Tiao, G. C. (1995). Modelling satellite ozone data, *J. Amer. Stat. Assoc.* **90**(431), pp. 969–983.

Northorp, P. (1998). A clustered spatial-temporal model of rainfall, *Proc. R. Soc. Lond. A* **454**, pp. 1875–1888.

Nott, D. J., and Dunsmuir, W. T. M. (2002). Estimation of nonstationary spatial covariance structure, *Biometrika* **89**(4), pp. 819–829.

Nualart, D. (1995). *The Malliavin Calculus and Related Topics*, Springer Verlag.

Oetoma, T., and Stevenson, M. (2005). Hot or cold? A comparison of different approaches to the pricing of weather derivatives, *J. Emerging Market Finance*, **4**(2), pp. 101–133.

Øksendal, B. (1998). *Stochastic Differential Equations. An Introduction with Applications*, Fifth Edition, Springer Verlag.

Onof, C., Faulkner, D., and Wheater, H. S. (1996). Design rainfall modelling in the Thames catchment, *J. Hydrol. Sciences* **41**(5), pp. 715–733.

Papazian, G. and Skiadopoulos, G. S. (2010). Modeling the dynamics of temperature with a view to weather derivatives, January 9, Available at SSRN: http://ssrn.com/abstract=1517293.

Park, M. S. and Heo, T.-Y. (2009). Seasonal spatial-temporal model for rainfall data of South Korea, *J. Applied Sciences Research*, **5**(5), pp. 565–572.

Pfeifer, P. E., and Deutsch, S. J. (1980). A three-stage iterative procedure for space-time modelling, *Technometrics* **22**, pp. 35–47.

Platen, E., and West, J. (2005). A fair pricing approach to weather derivatives, *Asian-Pac. Financial Markets* **11**(1), pp. 23–53.

Protter, Ph. (1990). *Stochastic Integration and Differential Equations*, Springer Verlag.

Rassmusson, E.M., Dickinson, R.E., Kutzback, J.E., and Cleveland, M.K. (1993). *Climatology*, in "Handbook on Hydrology", Maidment, D. R. (ed.), McGrew-Hill Professional.

Rehman, S., and Halawani, T. O. (1994). Statistical characteristics of wind in Saudi Arabia, *Renewable Energy*, **4**(8), pp. 949–956.

Ripley, B. D. (1981). *Spatial Statistics*, John Wiley, New York.

Rodriguez-Iturbe, I., and Mejia, A. (1974). The design of rainfall networks in time and space, *Water Res. Research*, **10**(4), pp. 713–728.

Rodriguez-Iturbe, I., Cox, D. R., and Isham, V. (1987). Some models for rainfall based on stochastic point processes, *Proc. R. Soc. Lond. A* **410**, pp. 269–288.

Rodriguez-Iturbe, I., Cox, D. R., and Isham, V. (1988). A point process model for rainfall: further developments, *Proc. R. Soc. Lond. A* **417**, pp. 283–298.

Roldan, J., and Woolhiser D. A. (1982). Stochastic daily precipitation models. 1. A comparison of occurrence processes, *Water Res. Research* **18**(5), pp. 1451–1459.

Rouhani, S., and Myers, D. E. (1990). Problems in space-time kriging of hydrological data, *Math. Geology* **22**(5), pp. 611–623.

Şahin, A. D., and Şen, Z. (2004). A new spatial prediction model and its application to wind records, *Theor. Appl. Climatology* **79**, pp. 45–54.

Šaltytė Benth, J., Benth, F. E., and Jalinskas, P. (2007). A spatial-temporal model for temperature with seasonal variance, *J. Appl. Stat.* **34**, pp. 823–841.

Šaltytė Benth, J., and Benth, F. E. (2010). Analysis and modelling of wind speed in New York, *J. Appl. Stat.* **37**, pp. 893–909.

Šaltytė Benth, J., and Benth, F. E.(2012). A critical view on temperature modelling in weather derivatives markets, *Energy Econ.* **34**, pp. 592–602.

Šaltytė Benth, J., and Šaltytė, L. (2011). Spatial-temporal model for wind speed in Lithuania, *J. Appl. Stat.* **38**, pp. 1151–1168.

Samuelson, P. (1965). Proof that properly anticipated prices fluctuate randomly, *Indust. Manag. Rev.*, **6**, pp. 41–44.

Sato, K. I. (1999). *Lévy Processes and Infinitely Divisible Distributions*, Cambridge University Press.

Schiller, F., Seidler, G., and Wimmer, M. (2010). Temperature models for pricing weather derivatives, *Quantit. Finance* **12**(3), pp. 489–500.

Schwartz, E. S. (1997). The stochastic behaviour of commodity prices: Implications for valuation and hedging, *J. Finance* **LII**(3), pp. 923–973.

Shiryaev, A. N. (1999). *Essentials of Stochastic Finance*, World Scientific, Singapore.

Skees, J. R. (2008). Innovations in index insurance for the poor in lower income countries, *Agricultural Res. Econ. Rev.*, **37**(1), pp. 1–15.

Snepvangers, J. J. J. C., Heuvelink, G. B. M., and Huisman, J. A. (2003). Soil water content interpolation using spatio-temporal kriging with external drift, *Geoderma*, **112**, pp. 253–271.

Solna, K., and Switzer, P. (1996). Time trend estimation for a geographical region, *J. Amer. Stat. Assoc.*, **91**(434), pp. 577–589.

Stein, M. L. (2005). Space-time covariance functions, *J. Amer. Stat. Assoc.*, **100**(469), pp. 310–321.

Stoll, S.-O., and Wiebauer, K. (2010). A spot price model for natural gas considering temperature as an exogenous factor and applications, *J. Energy Markets* **3**(3), pp. 113–128.

Svec, J. and Stevenson, M. (2007). Modelling and forecasting temperature based weather derivatives, *Global Finance J.* **18**, pp. 185–204.

Taib, C. M. I. C., and Benth, F. E. (2012). Pricing of temperature index insurance, *Rev. Develop. Finance* **2**(1), pp. 22–31.

Todini, E., and di Bacco, M. (1997). A combined Pólya process and mixture distribution approach to rainfall modelling, *Hydrol. Earth System Sci.* **1**(2), pp. 367–378.

Tol, R. S. J. (1996). Autoregressive conditional heteroscedasticity in daily temperature measurements, *Environmetrics* **7**, pp. 67–75.

Tol, R. S. J. (1997). Autoregressive conditional heteroscedasticity in daily wind speed measurements, *Theor. Appl. Climatology* **56**, pp. 113–122.

Torres, J. L., Garcia, A., De Blas, M., and De Francisco, A. (2005). Forecast of hourly average wind speed with ARMA models in Navarre (Spain), *Solar Energy* **79**, pp. 65–77.

Velarde, L. G. C., Migon, H. S., and Pereira, B. de B. (2004). Space-time modelling of rainfall data, *Environmetrics* **15**, pp. 561–576.

Wheater, H. S., Isham, V. S., Cox, D. R., Chandler, R. E., Kakou, A., Northorp, P. J., Oh, L., Onof, C., and Rodriguez-Iturbe, I. (2000). Spatial-temporal rainfall fields: modelling and statistical aspects, *Hydrol. Earth System Sci.* **4**(4), pp. 581–601.

Woolhiser D. A., and Roldan, J. (1982). Stochastic daily precipitation models. 2. A comparison of distribution of amounts, *Water Resources Res.* **18**(5), pp. 1461–1468.

Yan, Z., Bate, S., Chandler, R. E., Isham, V., and Wheather, H. (2002). An analysis of daily maximum wind speed in northwestern Europe using generalized linear models, *J. Climate* **15**, pp. 2073–2088.

Zapranis, A., and Alexandridis, A. (2008). Modelling the temperature time-dependent speed of mean reversion in the context of weather derivatives pricing, *Appl. Math. Finance* **15**(4), pp. 355–386.

Zhang, Z., and Switzer, P. (2007). Stochastic space-time regional rainfall modeling adapted to historical rain gauge data, *Water Resources Res.* **43**, W03441, 14 pp.

# Index